Osprey Modelling • 6

Modelling the SdKfz 251 Halftrack

Robert Oehler
Series editors Marcus Cowper and Nikolai Bogdanovic

First published in Great Britain in 2004 by Osprey Publishing,
Midland House, West Way, Botley, Oxford OX2 0PH, UK
443 Park Avenue South, New York, NY 10016, USA
Email: info@ospreypublishing.com

© 2004 Osprey Publishing Ltd.

All rights reserved. Apart from any fair dealing for the purpose of private study, research, criticism or review, as permitted under the Copyright, Designs and Patents Act, 1988, no part of this publication may be reproduced, stored in a retrieval system, or transmitted in any form or by any means, electronic, electrical, chemical, mechanical, optical, photocopying, recording or otherwise, without the prior written permission of the copyright owner. Enquiries should be addressed to the Publishers.

ISBN 1 84176 706 9

Consultant editor: Robert Oehler
Editorial by Ilios Publishing, Oxford, UK (www.iliospublishing.com)
Design: Servis Filmsetting Ltd, Manchester, UK
Index by Alison Worthington
Originated by Global Graphics, Prague, Czech Republic
Printed in China through Bookbuilders

06 07 08 09 10   11 10 9 8 7 6 5 4 3 2

A CIP catalog record for this book is available from the British Library.

FOR A CATALOG OF ALL BOOKS PUBLISHED BY OSPREY MILITARY AND AVIATION PLEASE CONTACT:

NORTH AMERICA
Osprey Direct, C/o Random House Distribution Center,
400 Hahn Road, Westminster, MD 21157
E-mail: info@ospreydirect.com

ALL OTHER REGIONS
Osprey Direct UK, P.O. Box 140, Wellingborough, Northants, NN8 2FA, UK
E-mail: info@ospreydirect.co.uk

www.ospreypublishing.com

Acknowledgements

I would like to express my gratitude to the following people who contributed a great deal to the completion of this title: My wife Valery for shouldering the extra burden throughout the course of this project. Special thanks for your patience, understanding and editorial assistance. Thank you, Valery, for making this book possible. You are incredible.
My assistant photographer (and seven-year-old daughter) Kendall Oehler for her help behind the camera and with her little brother Jake. To Jakob Oehler for being a good boy and allowing me the flexibility to pursue my hobby. My parents, for encouraging my early interest in the hobby.
In addition to my family, I would like to take this opportunity to thank the following individuals. Marcus Cowper and Nikolai Bogdanovic of Ilios Publishing, for your endless patience, advice and understanding. Kevin Kuster, for your friendship, assistance and support. Marcus Nicholls for being my all-time modeling hero (and source of much laughter) for many years. Karl Madcharo "The Dealer," supplier of all things model related. Thank you, my friend, for being an encyclopedic model and aftermarket resource. Charlie Pritchett for being a friend and modeling inspiration. Stan Spooner for your generosity and friendship, for your graciously positive personality and for your endless support.
The following individual manufacturers/distributors who contributed products to this project (specific contributions identified in the text and captions):
François Verlinden of Verlinden Productions
Bill Miley of Chesapeake Model Designs
Jon Tamkin of Mission Models
Woody Vondracek of Archer Fine Transfers
Vicki Leung of Dragon Models
Alasdair Johnston of The Small Shop EU and Gerard Dambrosio of Model Maker Products
John Smith of Silent Aire Technologies and Medea/Artool
Rich and Joy Sullivan of R&J Enterprises

Contents

Introduction   4
Vehicle development • Importance of the SdKfz 251 as a modeling subject
Some personal thoughts on modeling

Materials and tools   8
Airbrushes, compressors and airbrush paint • Finishing "style"
Paint and planning a weathering strategy • Color accuracy and scale effect

SdKfz 251/1 D in 1/72 scale   16
Decals • Application of pastels

SdKfz 251/1 C in 1/35 scale   21
Drives and suspension • Road wheels • The tracks • Upper hull • Hull interior • Fenders
Interior painting • A word about washes • Joining the upper and lower hull
Painting the machine guns • Application of mud

SdKfz 251/9 "Stummel" late   32
The gun mount • Piecing together the fighting compartment • Exterior components
Running gear • Painting and weathering • Joining the upper and lower hull • Oh yeah, the radio …
Exterior painting • Markings and insignia • Weathering • Applying mud • Chipping

SdKfz 251/7 engineer vehicle with assault bridge   46
Soldering • Running gear • Storage racks • The engine • Interior details • Exterior details
Painting and weathering • Markings • Tracks and tires • Joining the upper and lower hull
Weathering the underside of the model • Chipping

A gracious gift   66
Composing the diorama • Painting the base • The shrine
Standing water • Shade tree • The Kubelwagen and cart

Further reading, media, websites and muscums   77
Books • Web pages/magazines • Museums and collections in possession of the SdKfz 251

Kits available   78
Aftermarket and photo-etch sets • Tracks • Conversion sets

Index   80

Introduction

Anyone with even a remote familiarity with the beginning of World War II has heard of blitzkrieg or "Lightning War." This concept, introduced to the world by Hitler's Germany, is defined by coordinated and concentrated attacks by both air and land forces. In specific reference to land forces, the blitzkrieg battle doctrine requires that infantry move along with tanks to exploit and secure breaches opened by the punch of the armored force. In order for this doctrine to be tested and employed, it was recognized that infantry transport needed to be advanced from horses to another mode capable of keeping pace with the new fast-moving armor.

This need was filled by a new type of vehicle based on the Hanomag 3-ton chassis (SdKfz 11) and was developed into the Mittlerer Gepanzerter Mannschaftskraftwagen (medium armored transport vehicle) during the mid-thirties. The Sonderkraftfahrzeug (SdKfz) 251 or "Mittlerer Schutzenpanzer-wagen, (SPW)" as it became known, was one of the most numerous vehicles in the German arsenal during World War II.

Vehicle development

Between 1939 and 1945 the SdKfz 251 saw service on every front from Russia to France, spanning the entire length of the war. During this period, six different manufacturers produced a combined total of 15,200 vehicles. There were four basic production models that were mechanically similar with external detail differences, each being a further simplification of its predecessor reducing production time and costs (the greater majority of the total being the C and D body types).

Body types

The representative body types were known as Ausführungs (modifications) Ausf. A, B, C and D described below:

View from the rear of the Maybach HL 42 engine looking forward. From this angle one can see the manifold and exhaust system on the port side. On the right-hand side the carburetor can be seen. Farthest forward are the cooling fans, seated directly in front of the radiator.

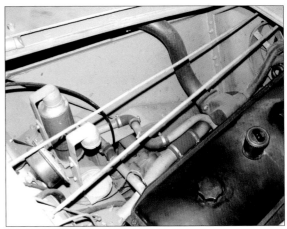

A closer view of the left-hand side reveals some of the plumbing details. This vehicle was restored with some components that were not typical of wartime vehicles, such as some of the wiring and the fire extinguisher cylinder evident in this view.

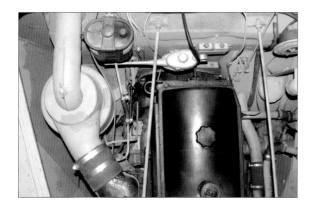

In this view, the air cleaner dominates the left-hand side of the compartment. The fluid reservoir is to its right, and an oil can as well as the Hella horn are on the right-hand side. The throttle and choke linkages can also be seen on the bulkhead between the oil can and the empty extinguisher bracket.

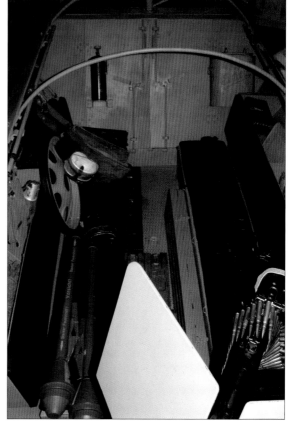

This view of the fighting compartment looking aft reveals lots of reenactor's gear. Panzerfausts, a rucksack, bread bag, meal tin, a spare road wheel, an MG42 and a number of ammo cans. The most interesting piece is the food-warming can marked with the word "Junkers" fixed to the left rear wall of the compartment.

Ausf. A: First production model built until 1940. This configuration featured three prominent vision ports on both upper sides of the hull superstructure. Its construction was complicated, featuring sloping armored plates that converged to form a two-part nose and tail. The engine had three armored doors surrounding the engine compartment and a grille immediately in front of the main engine access doors for cooling. The radio aerial was fitted to the right front fender and many of the pioneer tools were attached to the upper rear superstructure on both sides of the vehicle.

Ausf. B: Built until 1940. View ports in the rear body sides were eliminated, interior stowage was reorganized and pioneer tools were relocated onto the side fenders.

Ausf. C: The most obvious change was the new single-piece armored nose plate at the front of the vehicle. In this configuration, air was drawn up behind the lower front plate. The top grille was no longer needed and was, therefore, eliminated. Fixed side vent covers for improved air circulation around the engine replaced the side and front cooling doors. There were also interior changes to the seating and the internal stowage configuration. The side fenders were slightly raised overall and were also angled upward just behind the drive sprocket to allow more room for earth build-up in the running gear. Pioneer tools were moved to the front mudguards. The radio was moved under the armor in front of the passenger/radio operator's seat. Both electric welding and riveting were used in the construction of the hull, due to the fact that some armament firms had facilities for riveting but not welding.

Ausf. D: The air intakes were moved under the engine side armor and permanent stowage lockers replaced the side fenders and stowage boxes. Most prominently, the armored body was simplified with a reverse sloping rear end, which eliminated a complex hinge mechanism on the crew access doors.

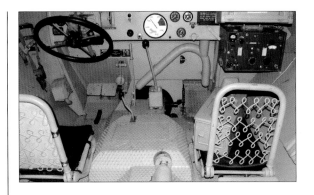 

The driver's and radio operator's compartment, while restored with non-standard items and incorrect radio equipment, gives a general view of the layout and cramped nature of this portion of the vehicle.

The radio operator's visor mechanism complete with armored glass and the head pad are shown to advantage in this view. The bracket in the center of the mechanism is for a forehead pad missing on this vehicle. Also, just to the right of the shot is the fixed vision block for the viewing port slit on the side of the compartment.

Mission-oriented configurations

In addition to the production variations, each body type was further divided into 23 different mission-oriented configurations.

SdKfz 251/1: mittlere Schützenpanzerwagen. Basic SPW fitted with the FuSpG "F" R/T unit

SdKfz 251/1 mittlere Schützenpanzerwagen (Mit Wurfrahmen 40). Nicknamed "Stuka zu fuss" (Stuka on foot)

SdKfz 251/2: mittlere Schützenpanzerwagen (Granatwerfer), Gerät 892. GrW34 mortar carrier

SdKfz 251/3: mittlere Funkpanzerwagen, Gerät 893. Communications vehicle, six sub-types

SdKfz 251/4: mittlere Schützenpanzerwagen (IG), Gerät 904. Light artillery tractor

SdKfz 251/5: mittlere Schützenpanzerwagen (PI), Gerät 905. Engineer vehicle for special assault engineers

SdKfz 251/6: mittlere Kommandopanzerwagen. Command vehicle (similar to /3 but with additional command equipment)

SdKfz 251/7: mittlere Pionierpanzerwagen, Gerät 907. Basic engineer vehicle

SdKfz 251/8: mittlere Krankenpanzerwagen, Gerät 908. Field ambulance

SdKfz 251/9: mittlere Schützenpanzerwagen (7.5cm), Gerät 909. 7.5cm KwK37 L/24 support vehicle

SdKfz 251/10: mittlere Schützenpanzerwagen (3.7cm PaK), Gerät 910. 3.7cm PaK36 light anti-tank gun vehicle

SdKfz 251/11: mittlere Fernsprechpanzerwagen, Gerät 911. Telephone communications vehicle (cable laying)

SdKfz 251/12: mittlere Messtrupp und Gerätpanzerwagen, Gerät 912. Artillery surveying vehicle

SdKfz 251/13: mittlere Schallaufnahmepanzerwagen, Gerät 913. Artillery surveying vehicle (Sound-recording)

SdKfz 251/14: mittlere Schallauswertepanzerwagen, Gerät 914. Artillery surveying vehicle (Sound-ranging)

SdKfz 251/15: mittlere Lichtauswertepanzerwagen, Gerät 915. Artillery surveying vehicle (Flash-spotting)

SdKfz 251/16: mittlere Flammpanzerwagen, Gerät 916. Flamethrower vehicle

SdKfz 251/17: mittlere Schützenpanzerwagen (2cm), Gerät 917. 2cm Flak38 antiaircraft vehicle

SdKfz 251/18: mittlere Beobachtungspanzerwagen, Gerät 918. Artillery observation and command vehicle

SdKfz 251/19: mittlere Fernsprechbetriebspanzerwagen, Gerät 919. Mobile telephone exchange vehicle

SdKfz 251/20: mittlere Schützenpanzerwagen (Infrarotscheinwerfer), Gerät 920. 60cm infrared "Uhu" searchlight vehicle (Ausf. D only)

SdKfz 251/21: mittlere Schützenpanzerwagen (Drilling MG151S), Gerät 921. Triple 2cm MG151 and 151/20 (drilling) antiaircraft vehicle

SdKfz 251/22: mittlere Schützenpanzerwagen (7.5cm PaK40), Gerät 922. 7.5cm Pak40 anti-tank vehicle

SdKfz 251/23: mittlere Schützenpanzerwagen (2cm KwK), Gerät 923. Reconnaissance vehicle (mounting 2cm KwK38 turret from SdKfz 234/1)

Importance of the SdKfz 251 as a modeling subject

Due to its sheer numbers, prolific appearance in every theater of the war, and its importance in Germany's battle doctrine, the SdKfz 251 is arguably one of the most important vehicles of the war. The variations in body design through its production life, its mission-oriented variations, and paint schemes provide the modeler almost endless options.

Unfortunately for 251 fans, aftermarket companies have almost ignored the subject (at least in comparison to the vast number of variants). Only within the past three to four years have we seen a significant attempt to improve available kits. While companies like Azimut, MR Models and Verlinden Productions have offered several conversion sets, none had really attempted any significant interior detail enhancements for the basic chassis until recently. Now Royal Model, Eduard, Part and R&J Enterprises have started a resurgence of interest in the vehicle with their recent offerings.

In addition to the statements above, the fact that it is an open-topped vehicle makes it the perfect subject for modeling in diorama and static display. After many years of waiting, manufacturers seem to be waking up to the endless possibilities available with the creation of a small number of well done chassis kits to serve as a basis for other kits representing the variations noted above. With the advent of newer kits, it will soon be difficult to list all of the aftermarket products that may become available. In the meantime, I have included the available sets that I have found most useful.

Some personal thoughts on modeling

Modeling is my art and chosen diversion. I am passionate about this hobby and remain very active helping others find the peace of mind that it brings to me.

The careful balance between family and work considerations limits the time available for this unique and worthwhile pursuit. What little time that is available to me is spent working in my workshop (every man needs a cave), building, painting or researching.

While I enjoy the process of researching the subjects, building and applying what I learn to the model, I am not a rivet counter obsessed with creating the perfect scale representation of each subject. On the other hand, I often dream of having the time to undertake the perfectly accurate, museum-quality, super-detail projects complete with full interiors that I see in magazines. I certainly respect and appreciate the work and craftsmanship that goes into these elaborate scratchbuilt masterpieces, but I am personally content using aftermarket items and my own details to improve the accuracy and detail of a good base kit. To sum it all up, I build for me and I build for fun.

Considering the high quality of available kits and aftermarket accessories these days, much of the work is already done for the modeler. With this in mind, the focus of this book is on four different projects of increasing complexity. Starting with an "out-of-the-box" model of a 1/72-scale SdKfz 251/1 D, moving on to an improvement of Dragon's SdKfz 251/1 C, a kit-bash conversion of Tamiya's SdKfz 251/1 D to the SdKfz 251/9 D "Stummel," and finally to a master level project aimed at improving and converting the Tamiya SdKfz 251/1 D to the SdKfz 251/7 D Pionerpanzerwagen. The special feature is a diorama representing a field photo op that I call "A fine day out."

Materials and tools

My father always says that "good tools make all the difference" and that "it's important to have the right tool for the job." Over the years, I have come to embrace the wisdom of these adages, relying on experience to understand the true usefulness and value of an item prior to purchase. With all of the appealing model gadgets on the market, it is all too easy for the inexperienced modeler to end up with a collection of once-used hobby tools that looked really useful at the time of purchase.

On the other hand, I know modelers who spend hundreds of dollars on high-end kits and then buy low-quality tools only to be dissatisfied with their end result. This is especially true when it comes to airbrushes and compressors.

Many plastic modelers build their first model with the same basic tools. Whether they are a young person trying to build the kit that some well-intentioned relative gave them for their birthday or an adult quietly trying to re-capture a portion of their misspent youth, many will start in their own quiet solitude, with a hobby knife, glue (sometimes tube glue), perhaps some sandpaper and the paint recommended on the kit packaging. If they are lucky, they may know a modeler nearby to ask questions. Unfortunately, most do not. So they struggle on, alone.

While a model can certainly be built using the tools and materials mentioned above, the end result is often frustration. If one survives this first modeling experience with the desire to try again, the undaunted individual may seek to buy a magazine or a book for assistance. They will strive to learn about tools and techniques in order to make the next project more enjoyable and the end result look more intentional and realistic in appearance.

In any case, the acquisition of items in our toolbox is often a process that parallels our skill development as modelers. For example, we learn early on that liquid cement is better than tube glue, that a certain sanding tool is more efficient, or that a side cutter is better than prying the part from its sprue with a hobby knife.

As part of this very individualized progression, some modelers will focus their attention on painting and others go beyond box modeling and move into super-detailing and scratchbuilding. Some will do both. Whatever the interest, their collection of tools and materials are the tangible examples of their experience. With this in mind, I have included a section of tools and materials that reflect my interest and experience. These are tools that I find most helpful and were used while building the models featured in this book. Hopefully, the suggestions will even save you some money (for new kits and books, of course).

Airbrushes, compressors and airbrush paint

I have been using an airbrush for nearly 20 years. During this time, I have tried almost every brush on the market. I currently own and use eight different brushes. What I've discovered in my search for the finest controlled spray pattern is the superiority of the dual-action airbrush with the matched headset design. The pre-eminent company that features this design is the Japanese company B.B. Rich. Brushes currently manufactured by B.B. Rich include the RichPen, Tamiya, Gunze Sangyo and Iwata airbrush lines. The matched headset design requires more precision in the machining process (thus, a little more expensive), but results in a needle/nozzle combination that offers the most control available—and that's the name of the game.

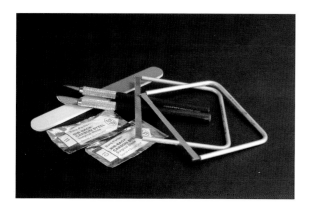

Cutting and sanding tools include Flexi-Files, inexpensive sanding sticks (found at the local grocery store) and surgical steel scalpel blades (number 10 and number 11) inserted into Tamiya modeling knife handles.

Tamiya extra thin cement flows extremely easily and is excellent to apply using the brush applicator. Tamiya extra thin also works to dilute Tamiya putty allowing the modeler to apply texture and, when appropriately thinned, works well as a surfacer.

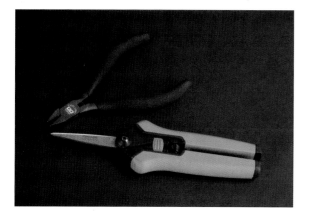

Tamiya side cutters are the best tools around for cutting plastic. They provide a clean, flush cut without pulling or creating unwanted holes in plastic. The inexpensive Fiskars Micro-Tip scissors are ideal for cutting wire and photo-etch due to their fine tip and precision cut.

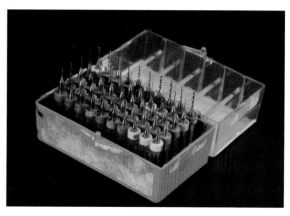

Carbon drill bits come in a variety of sizes and are very handy to buy in assortments like the one shown.

Photo-etch bending brakes have been some of the most unique and well-marketed tools in the past several years. Shown in the photo are the Etch Mate by Mission Models and the Hold and Fold by The Small Shop. Both are very versatile single knob designs and are excellent for larger items.

The dual knob design of 8in. Hold and Fold was particularly helpful in bending extremely small parts.

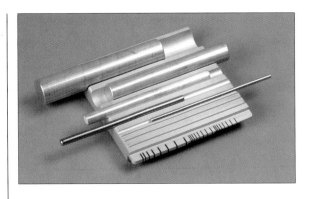

This clever item is the photo-etch rolling tool from the Small Shop. I found this item to be particularly helpful when creating the buckets.

Dental Ventures cyanoacrylate (CA) glue and accelerator. The design of the dispenser allows the user to squeeze out one drop at a time. The applicator never dries up due to the Teflon tip.

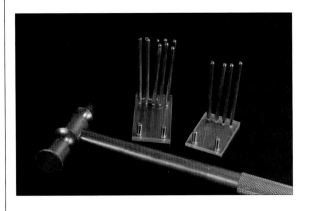

Punch and die sets (both round and hexagonal) and small brass hammer for creating bolt heads in various scales.

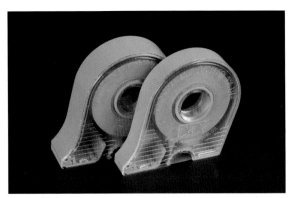

Always handy, Tamiya modeling tape is low-tack and comes in 6, 10, 18 and 25mm sizes. Great for everything from masking to holding small parts.

I will generally use the Iwata Revolution for area work and the Iwata Custom Micron-SB (CM-SB) for up-close detail work. Those who are familiar with my previous work know that I like to paint detail and weather with the airbrush. As a result, the Custom Micron is, by far, my favorite. Another plus is the placement of the color cup, which is off to the side rather than on top of the airbrush. With the CM-SB one can change the color cup from one side of the brush to the other to suit personal taste. Being left-handed this is a very useful option. This is especially helpful when completing close detail work, because I am able to look directly over the tip of the brush without a color cup obscuring my view.

Compressor

I use the Silent Aire, Super Silent ECO Air 20 because it encompasses the two features that I consider important in a compressor. First, it has an integral air tank. This feature is important because the air comes from the compressor in a steady flow. When your air is steady, so is your spray pattern. By contrast, the diaphragm compressor design that is sold in most hobby shops is unsteady. With the diaphragm design, the airflow can be described as somewhat of a "put-put-put." The result is a spay pattern that looks something like the following "- - - - -."

The second important feature of this compressor is that it is silent. I often watch movies or listen to the radio when I work, so the silence is much appreciated.

Arguably, the most important feature to be added to any compressor is the moisture trap/regulator combo, which is generally sold separately. This item

Fine wire is always helpful to have. The spool on the left is brass wire and was acquired by dismantling a Waterpik dental tool. The spool on the right is lead wire purchased from a sporting goods shop (used for fly fishing).

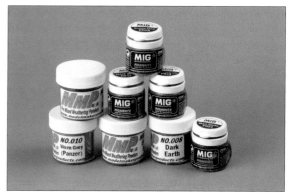

Pre-ground pastel pigments and powders are the cause for the recent weathering rage. Gone are the days where you have to crush or grind the pastel stick yourself! They come in a variety of colors and work well wet or dry.

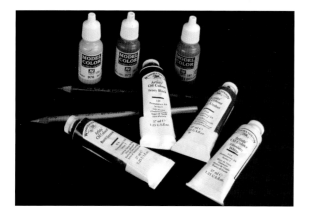

Vallejo acrylic paint is used for fine detailing, painting figures and outlining. Winsor & Newton oil paints are primarily used for washes and creating gun metal, along with Prismacolor art pencils.

Tamiya acrylics and acrylic thinner. In my estimate, the finest paint combination on the market.

allows the operator to control or regulate the pressure from the compressor to the airbrush. Stated in very general terms—the lower the pressure, the finer the line. The moisture trap does just what its namesake suggests; it traps moisture (condensation) in the line, preventing it from running through the airbrush and disrupting the airflow causing the paint to splatter and destroy your perfect paint job. I have also added a dual valve manifold so that I can independently operate two different airbrushes from the same air source.

Airbrush paint

Tamiya paints are reputed to be non-toxic and are specifically formulated for airbrush use. As a result, they enable the modeller to spray with maximum control at an air pressures as low as 10psi. I find the combination of Tamiya thinner and acrylic XF (flat) colors to be the perfect combination for airbrush work.

One of the relatively unknown advantages of using Tamiya acrylics is the quick clean-up using glass cleaner with ammonia. Glass cleaner with ammonia quickly dissolves even dried Tamiya acrylic paint. So, to clean your airbrush all you have to do is dump out the excess paint, pour some glass cleaner into the

Plasticote automotive primer for plastic and Testors' products: Dullcote (spray and bottle), and airbrush thinner for diluting the Dullcote for airbrush use.

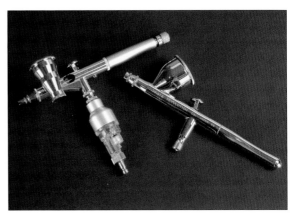

Weapons of choice: Iwata Custom Micron-SB and the Iwata Revolution, both superb airbrushes offering excellent control and good balance at an affordable price.

color cup and then spray it through the airbrush until the spray is clear in appearance. As a final step, pour a small amount of glass cleaner in the color cup. Then place your finger over the nozzle and apply the air by depressing (or pulling back, in the case of a dual-action airbrush) the trigger and "back-flow" the paint from the needle back into the color cup. Finally, remove your finger and spray out the remaining paint and glass cleaner from the color cup and that's it, you are finished! I have met people over the years that have stated that they don't use Tamiya acrylics because they have problems with "spattering" and the "sandy texture" that forms on the surface of the model. Fortunately, these two issues are both easy to correct.

The "spattering" happens because the paint is too thick and is building up on the tip of the needle. To keep this from happening you must do several things. First, clean your brush ensuring that the needle is clean and straight. A bent needle collects paint and misdirects airflow. When you mix your paint, try

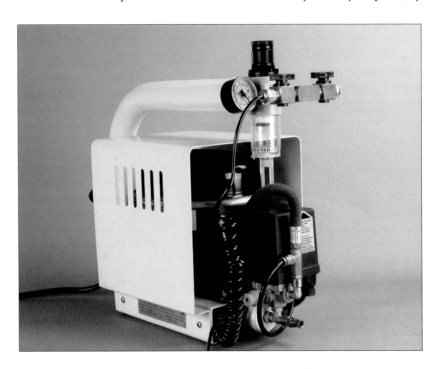

My favorite compressor! The Silent Aire, Super Silent ECO Air 20. So quiet you can work with the television on.

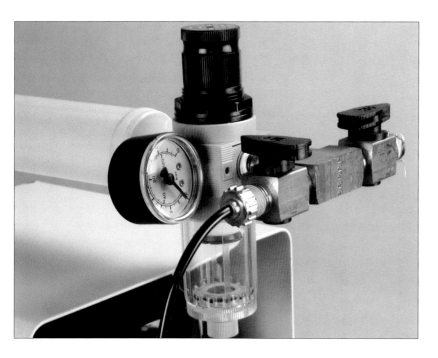

Filter regulator combination so that I can trap moisture and adjust pressure output. Attached is a dual manifold with valves so I can run both the Custom Micron-SB and the Revolution at the same time.

a paint to thinner ratio of 50/50, gradually adding more thinner until there is nothing but a fine line coming from the nozzle of your airbrush. I mix my paint in the color cup with a dropper so that I can add thinner and test the line in order to ensure a good mix every time.

The "sandy texture" occurs when the pressure is too high and the nozzle is too far from the subject. An airbrush is somewhat of an atomizer. Under pressure, the paint is atomized and shot through the nozzle. If the paint reaches the surface of the subject while it is wet, it flows smoothly onto the surface of the model. However, if the pressure is too high and the nozzle is too far away from the surface of the subject, the atomized paint turns into little microscopic rocks that "pile-up" creating the "sandy texture." To avoid this common problem, simply turn down your pressure and bring the nozzle closer to the model. This, of course, means that you'll have to work a smaller area. Perhaps it also means that some modelers will need to approach these issues with an added measure of patience.

Finishing "style"

There are many "styles" of weathering employed by the current generation of modelers. A few years back, one would paint the model one color, slap on the markings, dry brush, maybe apply a little groundwork and put the model up for display. Most recently, if you've re-entered the hobby, you've noticed a dramatic change in the finishing of armor models. Weathering, for example, now consists of highlighted finishes, metallizers, streaking and washes with oil paints, pastel powders, etc. Fortunately for us, this trend of more realistic weathering is relatively inexpensive and easy to learn.

If you see my finished work in person it is, admittedly, dark. This is due to how the model "presents itself" to the camera lens under photographic lighting conditions. Photographic lighting tends to make a model look flat and uninteresting. To counter this phenomenon, I have developed a style of finishing to compensate by adding more contrast to the facets of the model.

In real life the idea behind vehicular camouflage is to paint the subject in such a way that the paint hides the facets of the vehicle when viewed at a distance, thereby blending it into its intended background. Conversely, when one wants to display their modeling efforts through photography, it is

Not normally considered a tool, but one of the most useful items I've used in a long time. Archer Fine Transfers uniform patches are so easy to use, they make the thought of actually painting them impractical.

Probably the most important tools in the modeler's inventory—good references. One can never have too many references.

necessary to paint in accents and highlights so that all of the facets of the model can be seen and appreciated. Adding accents and highlights is a simple process that is demonstrated in the photographs.

Paint and planning a weathering strategy

Of all the combatants during World War II, the German military had the most developed vehicular camouflage. While all vehicles came from the factory with at least a basic paint finish, crews were issued concentrated paint paste for application in the field. This paste was cut with thinner (factory applied), water, gasoline, kerosene and, probably, whatever else was available in the field environment. Depending on the method of application (veterans' accounts suggest that the Germans applied their paint with everything from a brush or broom, to an air gun and even an old shirt), and the degree of dilution, it is clear that there are endless permutations of regulation and non-regulation schemes.

When planning a weathering strategy, it is important to review your references. A careful review of the crew access hatches and mounting points of a vehicle will tell you a great deal about how it was used and how it truly appeared under combat conditions. One should also consider the different layers of paint on the areas of the vehicle that would have been polished (by crew movement), scraped, scratched, shot or otherwise damaged.

You may also want to remember the super-resilient epoxy primer that the Germans used to undercoat their armor. This stuff was so resilient, that it is extremely rare to see a worn steel edge on any portion of a German vehicle.

The age of the vehicle is important, too. The weathering approach will be very different for a model representing a recently delivered, factory-fresh vehicle as compared to one that has seen a great deal of combat or been sitting in an outdoor museum for the past 60 years.

The most common fault I see in extreme weathering is in scale models based on an actual vehicle that the builder has photographed at an outdoor museum. After taking copious photographs of his subject, the modeler weathers his subject the same way it appears in the museum. Then, as a finishing touch, he places several German figures on the model with a plaque that displays the words "January 1945."

While the individual above might have accurately portrayed the vehicle for 2004, it is certainly not what the subject looked like in 1945! The point I am trying to make is that it is important to consider all of the items mentioned above when planning a weathering strategy.

Color accuracy and scale effect

Color accuracy is a highly technical and controversial subject for modelers. Some individuals will insist on absolute, factory-referenced paint chips or Federal Standard references, while others are simply content to use the paint directly from the jar/bottle. My personal preference is to consider accuracy and scale effect, but I am not a purist. Neither am I dogmatic regarding this subject, I am simply one who finds a degree of comfort in reference.

I suppose with that introduction, I should explain what scale effect is. First of all, it is important to understand full-scale color. Full-scale color generally refers to production paint, as it is applied to a full-scale or a one-to-one scale subject. Therefore, it follows that scale effect is how color changes in appearance as the paint is scaled with the subject.

British modeler and color expert Ian Huntley explains it as "color that is observed and measured over a scale distance." To your eye, a 1/35-scale tank viewed from roughly $10^1/_2$ft away is the same in appearance to that of a full-scale subject 35ft away due to distance and atmospheric interference. Similarly, a 1/72-scale model painted with full-scale colors would appear too dark and, therefore odd and unauthentic.

Since one cannot accurately employ full-scale colors on a scale model, how does one best replicate scale effect? Once again I refer you back to Ian Huntley. He has proposed a list of acceptable percentages of white to reduce colors at each scale. For example, at 1/72 one would add 15 percent white to fresh, full strength, factory-produced paint. In 1/35 scale one would add roughly 8 percent white.

This is, of course, an abbreviated introduction (or perhaps, reminder) regarding the concept of scale effect. It is inevitable that modelers will continue to debate the accuracy of color on a given subject. While this concept is an attempt to quantify the relationship between color and scale in reference to atmospheric perspective, it is far easier for me to say, "Color is subjective." So, my advice is to choose your colors in a way that is acceptable to your personal standards.

Some references on color:

Chory, Thomas, *Wehrmacht Heer Camouflage Colors, 1939–1945*, Aura design Studio, 2000

Klaus, David H., *The IPMS Cross-Reference Guide*, 1988

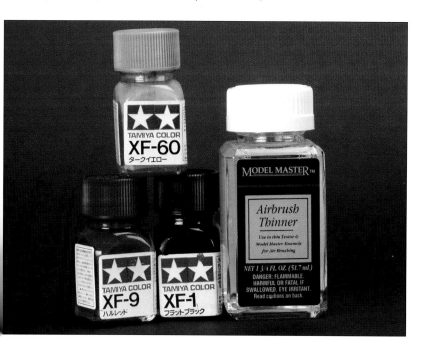

Generally used for simulating paint chips, I use Tamiya enamel paints along with the Testors' airbrush thinner. While not widely available outside of Japan, they are the finest enamel paints on the market. They are also convenient, because the paint colors match the Tamiya line of acrylic paints.

SdKfz 251/1 D in 1/72 scale

Subject:	SdKfz 251/1 D in 1/72 scale
Model by:	Robert Oehler
Skill level:	Intermediate
Basic kit:	Hasegawa SdKfz 251/1 D (item MT44)
Markings:	Kit decals
Paints:	Tamiya XF-60 Dark Yellow, XF-64 Red Brown, XF-57 Buff, Dullcote Laquer-Flat

When developing the synopsis for this book, I looked at creating a build effort that only included models in 1/35 scale. I've always enjoyed 1/72 scale, and have been very impressed with the new molding technology and the level of detail present in many of the newer kits. So, in this one case, I decided to forgo the 1/35 and take the 1/72 plunge.

The Hasagawa kit is an excellent representation of the late D with one exception, the tracks. Included in the kit are two runs of over-scale vinyl, or "rubber-band," tracks. Since the rest of the model is so well done, I hope some enterprising aftermarket company offers a set of appropriately scaled tracks.

The kit was basically built from the box, and as I went through the steps, I was amazed at the detail represented on such small parts.

This finishing strategy is the same basic method I used on the subsequent models featured in this book. Tamiya acrylic Red Brown was used as a primer and sub-components of the model received this application.

I then applied Tamiya acrylic Dark Yellow, XF-60, across the entire model. This was achieved by carefully highlighting the inner portions of each panel section, leaving a darker yellow/brown edge on the outside of each. When the highlighting was satisfactory, on both the interior and exterior, I began painting the camouflage.

When I was satisfied with the look of the dark yellow, I began laying on the randomly applied striped camouflage pattern with Tamiya Red Brown, XF-64.

Then I slightly lightened the red brown with buff and applied it as a highlight sprayed inside of the previously established red brown pattern.

Next, I detail-washed the entire model with Black Winsor & Newton oil paint heavily thinned with Turpenoid to pick out the recessed and raised features. Once again, this is done by placing the mixture on a liner brush next to (or

The Hasegawa SdKfz 251/1 D kit in 1/72 scale is a very small kit as compared to the dime in the photograph.

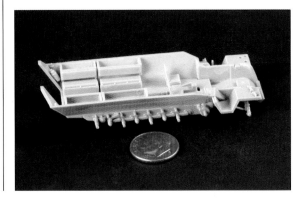
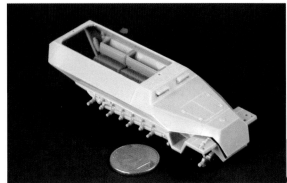

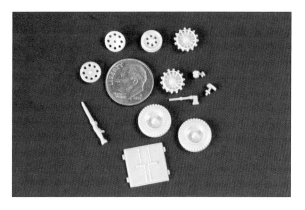

Another perspective view displaying all of the minuscule components in reference to the dime.

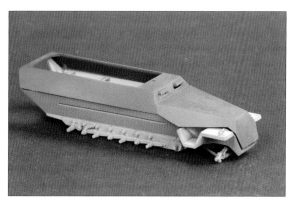

The model was easy to build straight from the box. Despite its size I found it to be a pleasure to build and paint. As described in other chapters of this book, I started the painting process by spraying the model with Tamiya acrylic Red Brown. The photograph shows that I highlighted the lower half of the model with Tamiya XF-60 Dark Yellow and left the top half of the model to demonstrate that the two were separated so that I could easily paint the interior of the model before cementing the upper and lower halves together.

The decals in the kit are very thick for this scale. To hide this problem it is necessary to coat the model with Future floor wax. When the wax is dry, the decal is placed onto the wax. The decal settles into the wax as it dries. Another coat is sprayed over the top of the decal as shown in the photo. When the final coat of wax dries, 0000 steel wool is used to lightly sand across the surface of the wax. When finished, the edge of the decal is almost invisible. For more details on this process see the chapter on the SdKfz 251/1 C.

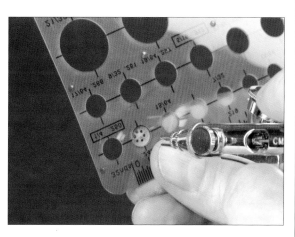

Due to the tiny size and number of the road wheels, I had to use an alternative method to paint them. The method chosen is displayed in the photo above. I simply sprayed each road wheel Tamiya XF-1 (Black). Then I placed the correct circle template diameter over the wheel and used the template as a mask. As a result, this task went from several hours to roughly 20 minutes.

into) the detail to be enhanced. Capillary action takes the paint around the detail without any additional assistance. Once the Turpenoid had evaporated, I dry-brushed the entire model using Winsor & Newton Yellow Ochre and Burnt Sienna. Each of these colors was lightly dry-brushed over the corresponding color, previously airbrushed onto the vehicle. After drying for a period of four hours, the model was sprayed with Testors' Dullcote. The Dullcote is an extremely important part of this painting style because it works to "flatten" all of the different layers of paint into one, adding a degree of depth to the look of the paint. When completely dry, the markings can be applied.

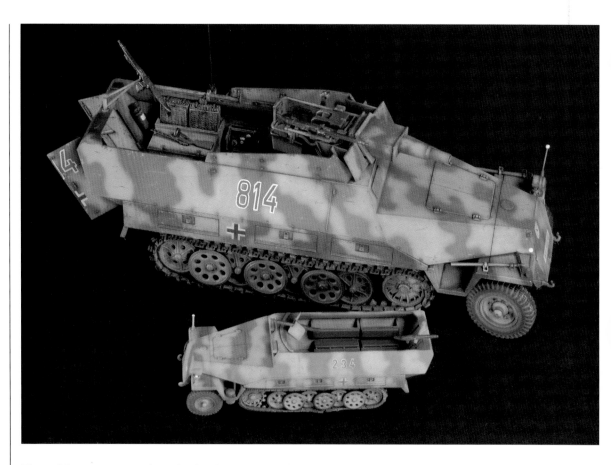

The model in proximity to its bigger brother. As you can see, there is a big difference in scale, but almost no difference in how one approaches the painting process.

Side views of the finished model. As one can readily see, the "rubber-band"-type tracks are too thick and grossly out of scale. This unfortunate item detracts from an otherwise superb little kit.

This overhead view of the front of the model reveals some of the weathering and paint detail as well as a glimpse into the fighting compartment.

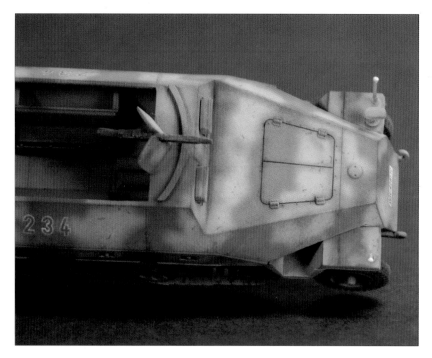

Decals

Due to the thickness of the decals in the kit, it was necessary to hide the edges by embedding the decal in clear acrylic floor wax (Future is my preference). I will generally begin this process by spraying the areas where the decals will sit with several layers of wax and allow the model to dry for about an hour. Then I carefully seat the water-slide decal in place, allow it to dry, and then spray another two coats of wax over the top of the decal. When completed, I allow the model to dry overnight.

The build up of wax (when done correctly) is almost invisible to the eye. However, to ensure that it is completely hidden, I sanded the top layers of wax using 0000 steel wool. It should be noted that this is somewhat of a delicate process and one should take care to not apply too much pressure in an effort to keep from exposing bare plastic.

I then added streaks and shadows with the airbrush. In the course of this process, it is important to remember that the markings must also be weathered. This is done by applying a highly diluted 5:2 mixture of Tamiya XF-2 Black and XF-10 Brown at approximately 12psi using the Iwata Custom Micron-SB. In order to further enhance the weathered look of the vehicle I also apply a detail wash using a Black/Burnt Sienna mixture of Winsor & Newton oil paints diluted with Turpenoid to all of the panel lines and raised details.

Application of pastels

The final stage of weathering is the application of pastels. At this point they are generally applied dry with a brush. I apply the pastel by lightly touching the model with a dollop of powder on the brush. The areas where the oil paints were deposited seem to attract and hold the powder. So when the excess is blown away, a gentle burnish with the tip of the brush will generally fix the pastel in place.

I also applied some "mud" chipping to the model, but I will save the description of these processes for one of the later chapters.

SdKfz 251/1 C in 1/35 scale

Subject:	SdKfz 251/1 C in 1/35 scale
Model by:	Robert Oehler
Skill level:	Intermediate
Basic kit:	Dragon Models Ltd. SdKfz 251 Ausf. C (item 6030)
Additional detailing sets used:	Seats from Royal Model update set number 086, Evergreen sheet plastic
Markings:	Kit decals
Paints:	Tamiya XF-69 NATO Black, XF-63 German Gray and XF-57 Buff, Dullcote Laquer-Flat

In this chapter we will look at some simple enhancements to the Dragon SdKfz 251/1 C. Overall, the Dragon offering is a very nice kit with excellent details such as the subtle weld seams and the highly detailed and workable tracks.

The hull consists of a shallow lower tub with separate torsion arms allowing the modeler to articulate the suspension as desired. On the inside of the tub are the transmission and drives, fuel tank, battery and oil tank. Due to the fact that many of these items cannot be seen when the upper hull is in place, I left them off to save time.

Drives and suspension

The first items that need improvement are the drive sprockets. Unfortunately Dragon did not include the small offset rollers that sit in between the halves of the drive sprocket. On the real vehicle, these rollers intermesh with the track horn to drive the vehicle forward.

To correct this problem I cut lengths of Evergreen rod and cemented them in position on the inner half of the drive sprocket. While not perfect, they serve to fill a void, providing a convincing enough alternative to what is provided in the kit. Once the cement dried, the pins were sanded to the appropriate height. Throughout the sanding process, careful repeated test-fitting ensured that each length of Evergreen rod properly sat against the outer drive sprocket. Once satisfied with the fit, both halves were cemented together using Tamiya extra thin cement.

Road wheels

A quick review of my references indicated that there is a slight gap between the steel wheel and the tire on the outermost road wheel. I corrected this visual deficiency by carefully deepening the gap rendered on the kit part with a pointed scriber. I then cleaned and scrubbed the scribed line with steel wool to remove any debris.

The tracks

The individual link tracks included in the kit are comparable to the SdKfz 251 Modelkasten tracks in both detail and function. Assembly was straightforward utilizing what has become the standard method for this type of track. Basically, each shoe is fitted together by locating pins, with the pad placed over the top of the pins thereby holding them in place.

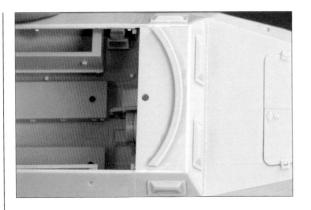

The hull roof displaying the Evergreen strip plastic shims I used to fill the open gap around the front and side areas of the roof panel. The shims were fixed with cyanoacrylate (CA) and the excess was sanded away. The result is a smooth surface that, when painted, will show no evidence of the former problem.

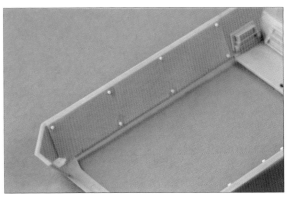

The underside of the upper hull provides a good view of the armored panel (integrally molded into the upper hull) and the bolt detail that I added. The bolts were formed from Evergreen sheet plastic using a punch and die set.

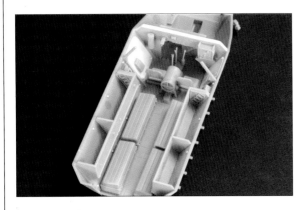

This interior detail of the lower hull displays the added details in white Evergreen plastic. Armored plate, shims on the control bulkhead (to ensure a snug fit with the upper hull) and bolt detail on the armored plate on the sides of the fighting compartment (out of view, obscured by the storage bins) complete the interior of the vehicle. The photo also reveals that I replaced the kit seats with resin seats found in the Royal Model detail set number RM-086.

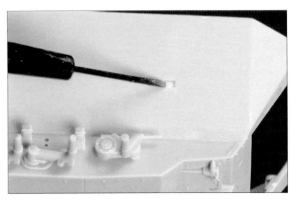

Another area that needed some attention was the locating holes for the fenders that went completely through the vehicle hull. To correct this problem, I fixed a piece of Evergreen strip into place with gap-filling CA and sanded both sides to a smooth finish. I then carefully chiseled shallower locator holes for the fenders.

Upper hull

The upper superstructure is a separate piece. Much of the interior "bolt-on" armor plate was integrally molded to the upper superstructure in the fighting compartment so I simply added the missing bolt detail. The bolts were formed from Evergreen sheet plastic using a Historex punch and die set.

Additional improvements consisted of shims made from Evergreen strip styrene on the control bulkhead (to fill the gap and ensure a snug fit with the upper hull) and the bolt detail that resides on the armored plate of both sides of the driver's/radio operator's compartment. I also replaced the kit seats with resin seats found in the Royal Model detail set number 086. The front of the driver's compartment has two-piece vision ports, consisting of an outer cast vision block and an inner housing containing the "armored glass."

When the hull roof was put in place, it was evident that some filling was required. This was done by placing strip styrene as shims in the gaps around

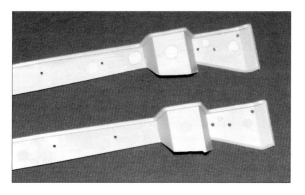

A before and after comparison of the fenders reveals that they also received a little attention. The finished fender on top shows the filled ejector pin marks and the shortened locator pins that were sanded almost flat so they could be received into the more shallow locating holes (of course, test-fitting is necessary during the sanding process to ensure a good fit).

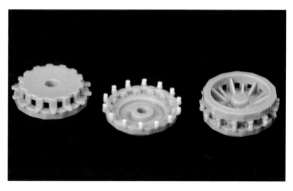

Another minor problem with the kit was the missing rollers that sit between the halves of the drive sprocket. To correct this problem, I cut lengths of Evergreen rod and cemented them in position on the inner half of the drive sprocket. Once the cement dried, the pins were sanded flat while carefully test-fitting them against the outer drive sprocket.

A quick review of my references indicated that there is a slight gap between the steel wheel and the tire on the outermost roadwheel. I corrected this visual deficiency by carefully deepening the gap rendered on the kit part with a pointed scriber. I then cleaned the scribed area with steel wool to remove any debris.

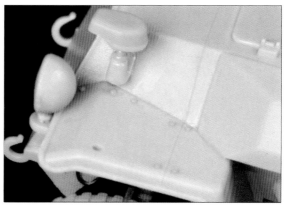

This photo reveals the bolt detail added to the fender area. The bolts were cut from the bottom hull of a Tamiya PzKpfw IV hull using a number 10 (curved) scalpel blade and fixed in place with Tamiya liquid cement.

the front and side areas of the roof panel. These shims were fixed with cyano-acrylate (CA) glue and the excess was sanded away. The result was a smooth surface that, when painted, showed no evidence of the former problem.

Hull interior

In addition to the previously mentioned seats and bolt-on armor, I added head pads to the underside of the hull roof. I chose to leave the rather large water tank, centrally located between the driver and radio operator, in place due to the fact that I would have to replace the floor plate in this area after filling the rather large locating holes. I believe the water tank was only featured on the 251/8 field ambulance, but I don't have any solid photographic evidence to prove or disprove this opinion.

In the fighting compartment, the troop seats sit on top of flat storage boxes. The seat backs are formed by the pads on the open, side panel, storage bins, which contain nicely rendered rifle racks.

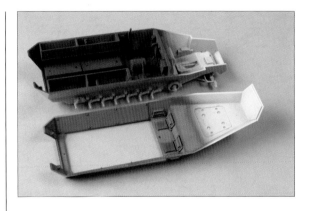

With the detailing and sub-assemblies complete, the painting process begins with the fighting compartment. I initially painted the entire inner portion of the vehicle with Tamiya XF-59 NATO Black. Then, as the photo indicates, I highlighted the interior with a mixture of Tamiya German Gray and Buff (see the mix recipe at the back of the book).

Driver's and radio operator's compartment. This area is blocked-in, highlighted and awaiting detail painting. The most unfortunate part of this kit is the large water tank between the driver's and radio operator's seats. This was a feature only seen on the ambulance variant. Due to the prominent locating holes in the anti-slip floor plates and the limited time I had to build the kit, I decided to leave the tank in place.

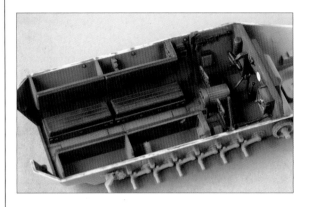

Final view of the lower fighting compartment prior to "buttoning it up."

Due to past experience with the seams of major assemblies cracking or splitting under stress (even with high quality CA), I usually turn to Tamiya extra thin liquid cement. To ensure a good weld, I brush apply cement to soften each side of the area to be joined. I then press the halves together, gently squeezing a bead of cement and softened plastic from the join line. The final step in this process is to secure the weld while it cures. I will usually secure the seam with tape as shown in the photo.

All major sub-assemblies complete. Ready for exterior painting.

Step one: Spray the entire exterior with Tamiya XF-59 NATO Black.

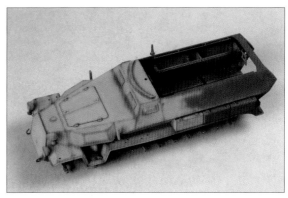

Step two: Using the same mix recipe previously mentioned, I highlight the inside of each panel surface on the entire model.

The model with completed highlighting.

Step three: In order to "flatten" the multiple layers of paint, I spray the entire model with Testors' Dullcote. An additional advantage to this step is that it also leaves an acceptable surface for decal and dry transfers.

Fenders

Another area that needed some attention was the locating holes for the fenders that went completely through the vehicle hull. To correct the problem here, I fixed a piece of Evergreen strip into place with gap-filling cyanoacrylate (CA) glue and sanded both sides to a smooth finish. I then carefully chiseled shallower locator holes for the fenders. The shortened locator pins were sanded almost flat so they could be received into the shallower holes. Regular test-fitting was necessary during the sanding process to ensure a good fit.

Additional bolt detail was also added to the front fender area. Feeling a little lazy and wanting to show another technique, I cut the bolts from the bottom hull of an old Tamiya PzKpfw IV hull using a number 10 (curved) scalpel blade and fixed them in place with Tamiya liquid cement.

The final enhancement was the wiring for the headlights mounted on the fenders. The conduit for these lights was simulated using solder glued in place with CA glue.

Interior painting

With the detailing and sub-assemblies complete, the painting process began inside the fighting compartment. As you can see by the photos, the lower and upper hull parts were left unattached to ease the painting of the interior.

Step four: Due to the thickness of the decals in the kit, it was necessary to hide the edges by embedding the decal in clear acrylic floor wax (Future is my preference). I will generally begin this process by spraying the areas where the decals will sit with several layers of wax and allow the model to dry for about an hour. Then I carefully seat the water-slide decal in place, allow it to dry and then spray another two coats of wax over the top of the decal. When completed, I allow the model to dry overnight.

I initially sprayed the entire inner portion of the fighting compartment with Tamiya XF-69, NATO Black. When it was dry, I then highlighted the entire interior with a mixture of Tamiya German Gray and Buff (See the mix recipe on the color samples at the rear of the book). The interior details were then blocked in and detail painted. When these items were dry, I sprayed the entire interior with Testors' Dullcote. When it is dry the Dullcote leaves an excellent semi-gloss finish, which is perfect for decals and oil washes. A detail wash of Winsor & Newton Ivory Black and Turpenoid was then applied using 10/0 liner brush. Once the wash was dry, Dullcote was applied by airbrush to slightly flatten the interior surfaces and details.

A word about washes

I use two basic types of washes:

A **detail wash** is typically applied using a finely pointed brush on a dry surface or one slightly dampened with Turpenoid. The oil paint and Turpenoid when mixed are the consistency of water and are only slightly transparent. This mixture is applied with the brush directly onto the individual detail you wish to highlight, such as a bolt or a length of panel line. Capillary action takes the paint around the detail without any additional assistance.

A **traditional wash** is applied using a large flat or watercolor brush over the entire surface of the model. The oil paint to Turpenoid ratio is very transparent as compared to the detail wash.

Joining the upper and lower hull

With the interior painting complete the upper and lower hull can be joined. It is important to note that the rear crew door hinges must also be fitted at this stage. Care must be taken to ensure that the hinges are properly aligned so the doors close properly. Due to past experience with the seams of major assemblies cracking or splitting under stress, even with high quality CA, I usually turn to Tamiya extra thin liquid cement. To ensure a good weld, I brush apply cement to soften each side of the area to be joined. I then press the halves together, gently squeezing a bead of cement and softened plastic from the join line. The final step in this process is to secure the weld while it cures. I will usually secure the seam with tape as shown in the photo.

Painting the machine guns

I airbrushed a black-gray finish onto the kit MG34 machine guns. Next a light tan dry-brush mixture of Winsor & Newton Burnt Sienna and White oil paints was created. After carefully rubbing the paint off of the brush onto a paper towel, I dipped the brush into a small bit of S&J Spray Metal polishing powder (aluminum). I rubbed the brush on a paper towel once again until it was almost clean. Then, I carefully dry-brushed the mixture onto the guns. With the dry-brushing complete, the S&J Spray Metal Polishing powder was rubbed in with a mixture of pastel and graphite. Once the desired sheen was achieved, further enhancements were made using a steel Prismacolor pencil. Finally the detail was further enhanced by a little light pastel work next to all of the raised detail.

Application of mud

See the chapter on the SdKfz 251/9 "Stummel" for the photo sequence and the text description of this process.

The build up of wax (when done correctly) is almost invisible to the eye. However, to ensure that it is completely hidden, I sand the top layers of wax down using 0000 steel wool. It should be noted that this is somewhat of a delicate process and one should take care to no apply too much pressure.

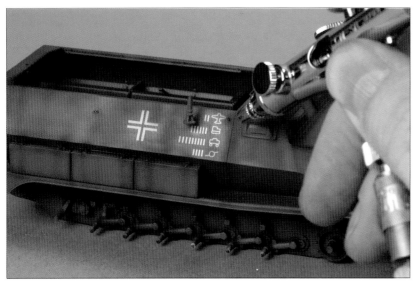

LEFT AND BELOW I add streaks and shadows with the airbrush. In the course of this process, it is important to remember that the markings must also be weathered. This is done by applying a highly diluted 5:2 mixture of Tamiya XF-2 Black and XF-10 Brown at approximately 12psi using the Iwata Custom Micron-SB. In order to further enhance the weathered look of the vehicle I also apply a detail wash using a Black/Burnt Sienna mixture of Winsor & Newton oil paints diluted with Turpenoid to all of the panel lines and raised details.

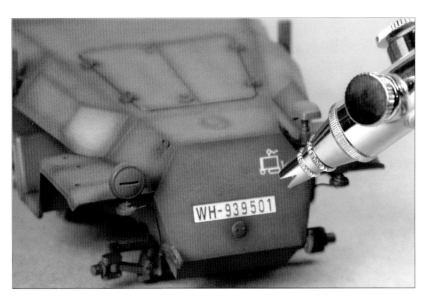

When the detail wash is dry, the entire model is lightly fogged with Testors' Dullcote. However, this time it is applied by airbrush to ensure a true flat finish across the entire surface of the model.

The final stage of weathering is the application of pastels. These are generally applied dry with a brush. I will apply the pastel by touching the model with a dollop of power on the brush. The areas where the oil paints have been applied seem to attract and hold the powder. So when the excess is blown away, a gentle burnish with the tip of the brush will fix the pastel in place.

One of the nice features of the kit is the workable hinges for the rear doors. When aligned properly, during the construction process, they will open, enhancing realism and interest.

Overall shot of the painted model prior to the installation of the tracks. Additional pastel work is done after the track are installed to avoid smudges and fingerprints.

Due to the unfortunate "thick look" of the jerry can holders that came in the kit, I chose to add the Royal Model equivalent. These elements were soldered for extra strength and applied to the model using CA cement. The cans themselves are from Tasca/Bego.

The spare tire was from the AFV Club wheels and tracks set and was created by using two of the same wheel halves. When glued together, they form a convincing hollow spare.

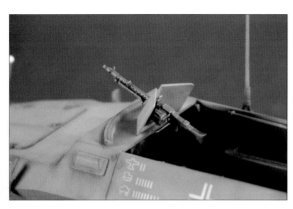

The kit MG34 machine guns provided at two stations on the vehicle are well rendered. They are finshed by rubbing S&J Spray Metal Polishing powder and graphite over a surface of black-gray.

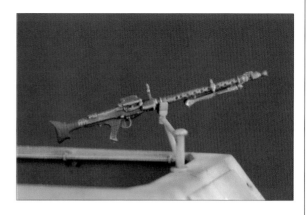

They are further enhanced by a little light pastel work next to all of the raised detail. In addition, the Prismaclor "steel" pencil is used to highlight some of the more polished portions of the gun.

PAGES 30–31 Multiple views of the finished model.

SdKfz 251/9 "Stummel" late

Subject:	SdKfz 251/9 "Stummel" late
Model by:	Robert Oehler
Skill level:	Advanced
Basic kit:	Tamiya SdKfz 251/1 Ausf. D (item 35195)
Additional detailing sets:	Dragon Models (item 6102) SdKfz 250/8 "Stummel," Plus Models metal buckets and cans (item 152), Jordi Rubio (item AC-01) German mudguard clearnace poles, Evergreen sheet plastic, Modelkasten SdKfz 251 tracks
Markings:	Eduard (item XT-049) SdKfz 251 markings, etched (item XT002), German crosses 2, (XT-002) late war
Paints:	Tamiya XF-60 Dark Yellow, XF-64 Red Brown, XF-58 Olive Green, XF-10 Brown, XF-52 Flat Earth, XF-57 Buff, Tamiya Enamels XF-60 Dark Yellow, XF-9 Hull Red, XF-1 Black, Dullcote Laquer-Flat

The Tamiya SdKfz 251/1 D (early) came out in 1990. At this time, like most Tamiya kits, it represented the best technological representation of this subject by any manufacturer. Even after all these years, it still holds up very well.

One small exception is the vinyl "rubber-band" tracks that come with the kit. I believe their use renders a well-made model appear toy-like. Even though some modelers are not fond of building individual link tracks, the addition of individual aftermarket, tracks result in a better, more realistic model. Modelkasten (Item K-019) tracks were my choice for this project.

The gun mount

After building the Dragon SdKfz 250 "Stummel" a few years back, I noted that the gun mount fitted very well onto the Tamiya SdKfz 251 and decided that it would be a fun way to get a late 251/9 from the Tamiya kit without too much pain and suffering.

However, after a little test-fitting, I decided the gun mount needed some assistance from Evergreen sheet styrene. Due to the thick nature of the baffled armor plates, I felt it necessary to replace the kit pieces with styrene sheet plastic. I also added the missing plate behind the gun within the recoil guard using sheet styrene and bolts made with the Historex punch and die set.

The upper side armor panel dimensions were derived using parts from the Dragon kit. They were physically laid out on Evergreen sheet styrene and cut to shape in an effort to achieve the desired scale thickness. The support brackets for the armored panels were made in the same manner. Most of the assembly was done in place on the upper hull. I also fabricated and installed the covered or blanked visor port on the starboard side along with the head pads and other brackets.

Piecing together the fighting compartment

The front section of the compartment floor was from the Tamiya SdKfz 251/1 D kit and the rear section with ammo locker was taken from the SdKfz 251/9 D (early) kit. The center section was carefully measured and cut to shape. The bottom of the styrene section was carefully sanded, and repeatedly test-fitted, to account for the thickness of the Eduard tread plate, which was also carefully

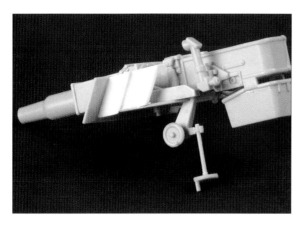

The 7.5cm KwK 37 (L/24) gun from the Dragon "Stummel" kit (item 6102) with styrene improvements. Due to the thick nature of the baffled armor, I felt it necessary to replace the kit pieces with styrene sheet plastic.

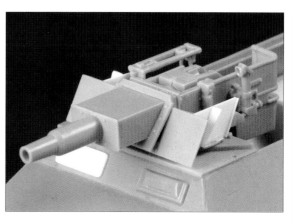

As you can see from the photo, the mounting for the gun in the Dragon Stummel kit is dimensionally very close to the roof of the Tamiya SdKfz 251/1 D kit, making it a fairly obvious conversion choice.

The missing plate behind the gun within the recoil guard was added using sheet styrene and bolts made from the punch and die set.

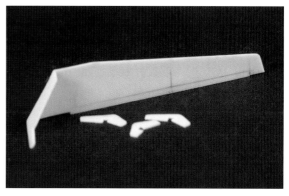

The upper side armor dimensions were derived using kit parts from the Dragon kit. Next, they were drawn on to Evergreen sheet styrene and cut to shape in an effort to achieve the desired scale thickness. The smaller items are the support brackets for the armor.

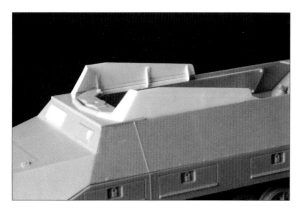

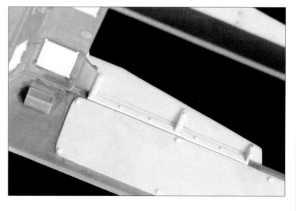

ABOVE AND ABOVE RIGHT Two views of the upper armor installed on the model. The upper view shows the outer view of the armor and its supporting brackets as well as the covered visor port that was blanked on these vehicles. The lower view shows the armor (both upper and interior), brackets, head pads and the blanked vision port to advantage.

cut to shape using the Fiskars Micro-shears. When all portions were adjusted to sit at the same height, the entire assembly was glued together.

To enhance the detail in the driver's compartment, I added a little styrene/brass detail to simulate the clutch, brake and throttle pedals. I also scratchbuilt the radio operator's folding seat and the mounting bracket for the FuG 5 radio. The radio itself is from the Tamiya SdKfz 251/1 D base kit and the wiring is solder.

I used the Mission Models Etch Mate to fold the ready-round rack for the 7.5cm rounds. A unique feature of the rack, unseen in the photos, are the clips that hold each round in place from behind. The entire rack and the clips are soldered for added strength. The beautiful turned-aluminum 7.5cm rounds are from New Connection.

I added the bolt-on armor sections on both sides of the driver's/radio operator's position as well as the the fighting compartment using Evergreen sheet styrene. I also included resin ammo cans from Plus Models and other interior items such as vision blocks, flare cases, the "Schmiesser" and the ammunition for the MP40. Each of these items was liberated from the Dragon Stummel kit.

The kit was not originally designed to accommodate interior components, so it is very important to check your work in regard to the fit of the upper and lower hull components. On several occasions, test-fitting revealed that the originally intended location for some of the parts had to be adjusted to ensure a proper fit.

Exterior components

In order to give the muffler a little more of a rusted and weathered look, I coated it with a thin layer of baking powder dampened with Tamiya extra thin cement. When painted and weathered, the baking powder gives the item a convincing texture.

Additional exterior features that were added consisted of the antenna mount and transformer, taken from the Royal Model set; the electrical conduit to the Notek light; and the tie down loops for the tarp that covered the fighting compartment in inclement weather.

Running gear

With the exception of the tracks, the only addition to the running gear was the spacer, provided by Modelkasten, that sits between both halves of the center road wheels. This spacer is an adaptive measure adopted to allow the road wheel to properly set over the guide horns on the track.

Painting and weathering

When painting a model with an interior I generally leave the top unfixed so that I can get at the interior details. Even though the 251 is an open-topped model, there is a great deal of detail that would be unreachable if I closed it up. To begin, I sprayed the entire model with Tamiya Hull Red XF-9, mixed with a few drops of Red Brown XF-64 (a ratio of 60:40 paint to thinner). I use this mix because it is the closest in tone to the primer used by the Germans.

Once the "primer" was dry, I followed by spraying Dark Yellow XF-60 on the inner portions of each section or panel, carefully leaving a darker yellow/brown edge. When finished, this panel highlight look presents an overall dark yellow vehicle with darkened edges at the panel lines. This method was utilized on both the interior and exterior of the model.

Interior painting

When the dark yellow was dry, I proceeded to block-in the interior details such as the seat pads, instruments, radio, etc. using Tamiya enamels by brush. These items were then weathered/highlighted with Vallejo acrylics.

The 7.5cm ready rounds were sprayed using metallic brass enamel that I bought at a craft shop. The tops of the rounds were then hand painted with black

Vallejo acrylic and the fuses were painted silver. Tamiya Clear Smoke was then sprayed around the base of each round in order to achieve a weathered look.

The gun itself, as well as the MG42, was painted by laying down a base coat of Tamiya XF-1 Black and then dry-brushing a mixture of Burnt Sienna, Black and Titanium White Winsor & Newton oils as well as S&J Spray Metal polishing powder over the surface. This mixture was then buffed with black dry pastel. A blued effect can be applied using Tamiya clear blue.

Painting accessories

The rolled tarp was used by the crew in inclement weather and was kept neatly folded in the starboard storage bin. This extremely well-sculpted item is from the Japanese firm Decal Star. Fortunately, it was almost the perfect size for the bin. After a bit of careful shaping, the part conformed neatly to the inside of its new home. This item was then painted with a darkened khaki/ flat earth mixture and a slightly lightened color was applied by airbrush as a highlight. The item was outlined using Vallejo acrylic Black and the strings were painted German tan.

The rear end of the vehicle was the perfect final resting place for the superb Plus Models bucket. A perusal of historical photos reveals that a bucket hanging on the exterior of a vehicle was quite a common practice. A couple of dents were added for realism.

One of the more impressive accessories is the water can by Bego. The cross was made by masking with Tamiya tape. The chipping effect was done as described later in this chapter.

Joining the upper and lower hull

With the interior painting complete the upper and lower hull can be joined. Tamiya extra thin liquid cement was used to ensure a good weld, the cement was brush-applied to both sides of the joint to soften the plastic. I then pressed the halves together by gently squeezing a bead of cement and softened plastic from the join line. The final step in this process was to secure the weld with tape while it cures.

Oh yeah, the radio ...

The radio was wired after I painted the interior and glued the upper and lower hull together. Even though I was very careful, portions of the interior needed to be re-painted. Luckily, I hadn't weathered the interior at this stage so there weren't any major disaster to be concerned with. The headphone connection was left open for the radioman crew figure that was later "plugged in."

Exterior painting

Before I laid in the camouflage pattern, I had to resolve a reference issue. After finding several different sources, I noted that there were some conflicting color references for vehicle 814 of the 20th Panzer Division. Not having a lot of time to consult the oracles and accurately resolve the problem, I decided to move forward with my own color interpretation of black and white historical photos. I decided that the vehicle was Dunkelgelb and Olivgrun. In any case I laid in a darkened version of Tamiya XF-58 (see color chips), and then highlighted the pattern using XF-58 lighted with XF-57 Buff.

After drying for a period of four hours the model was sprayed with Testors' Dullcote. The Dullcote is an extremely important part of this painting style because it works to "flatten" all of the different layers of paint into one layer adding a degree of depth to the look of the paint. When completely dry, the markings can be applied.

Markings and insignia

The next step in the process was the application of markings and insignia. I used three different sets from Eduard.

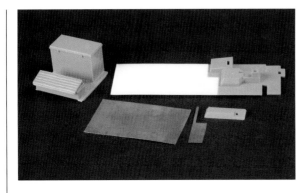 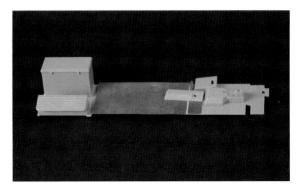

ABOVE AND ABOVE RIGHT These two views show the separate components of the complete hull floor. The first view shows the components disassembled. The front section was from the Tamiya SdKfz 251/1 D kit and the rear section with ammo locker was taken from the SdKfz 251/9 D early kit. The center section was carefully measured and cut to shape. The bottom of the styrene section was carefully sanded to account for the thickness of the Eduard tread plate, which was also carefully cut to shape using the Fiskars Micro-shears. The second photo (above right) shows all of the components test-fitted together prior to installation.

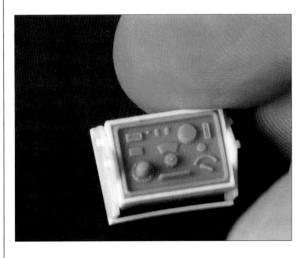 

The comparative size of my fingers indicates the size of the FuG 5 radio and scratchbuilt mounting bracket. The radio itself is from the Tamiya SdKfz 251/1 D base kit.

This unpainted display of the ready-round rack features soldered Eduard photo-etch and the 7.5cm rounds are from New Connection.

The FuG 5 radio shown mounted in position next to the scratchbuilt radio operator's folding seat.

To enhance the detail in the driver's compartment, I added a little styrene/brass detail to simulate, from left to right, the clutch, brake and throttle pedals.

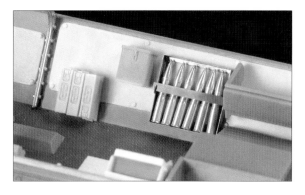

This unpainted view of the right-hand side of the model shows all of the components to advantage. The resin ammo cans are from Plus Models and the items shown in gray were procured from the Dragon Stummel kit.

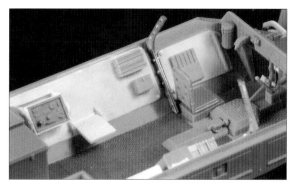

The unpainted view of the left-hand side of the model. Gray items include spare vision blocks, flare case and ammunition for the MP40. The "Schmiesser" was added later.

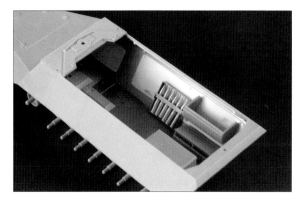

This somewhat out of sequence photo of the right-side interior with the upper hull in place is included to emphasize the need for constant test-fitting. It is often the case that modelers place mixed media aftermarket components into the lower part of the vehicle, and later find that the upper portion of the model doesn't seat properly. Also, this photo is a good view of the KwK 37 (L/24) gun mount.

In order to give the muffler a little more of a rusted and weathered look, I coated it with a thin layer of baking powder. When painted and weathered, the baking powder gives the item a convincing weather-beaten texture.

As shown in the photo, the first step is to paint the entire model with primer or hull red (mixed with a little red brown) to simulate the German primer color.

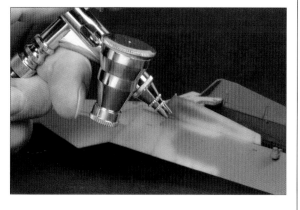

In the next step, Tamiya XF-60 Dark Yellow is used as a highlight over the primer color. As you can see in the photo, I visualize the model as a series of panels and then I paint the interior of each panel dark yellow.

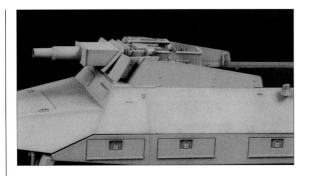

With the highlighting completed, the effect is clear. Each panel line stands out as a darker, somewhat reddish yellow. Later, additional painted shading, washes and pastel work will make these panel lines stand out even more.

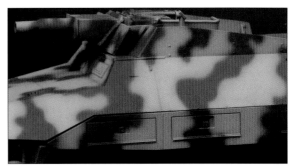

Due to conflicting color references from different sources for vehicle 814 of the 20th Panzer Division, I decided to move forward with my own color interpretation from black and white historical photos. In my interpretation, I decided that the vehicle was Dunkelgelb and Olivgrun. In any case I laid in a darkened version of Tamiya XF-58 (see color chips), and then highlighted the pattern using XF-58 lightened with XF-57 Buff.

The next step in the process is the application of markings and insignia I used three different sets from Eduard. For the numbers I used German numbers, medium (item XT008), German crosses late (item XT 002) and for the unit markings I chose the SdKfz 251 set (item XT 049). As you can see from the photo, I carefully applied the numbers to the model taking care to ensure they were straight. For reference, I used the additional vinyl pieces to maintain my alignment.

Step two: I sprayed the stencil mask with Tamiya XF-2 (Flat White) and, once dry, removed the stencil. Note the alignment bars are still in place.

Step three: Using the alignment bars as a guide, I applied the second, smaller mask over the previously applied white numerals. Finally, Tamiya XF-7 (Flat Red) was sprayed onto the mask.

When the red was dry, the mask was removed and the result is what you see before you. The cross seen below, between the doors on the sponson locker, was painted using Eduard express mask XT 002 in the same manner as described in the application of the numerals.

Left side view of the model, sporting the markings prior to weathering. One should also note here that the wheels are not yet glued in place in order to maintain access to the wheels and sponsons for the weathering effort yet to come.

The unfinished resin pieces are shown fitted to their positions. Once again, careful test-fitting is crucial when sanding/shaping the aftermarket items to naturally conform to their intended place. These extremely well sculpted and cast items are from the Japanese firm Decal Star.

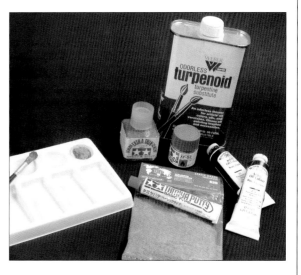

The application of markings to the crew access doors was as described previously. However, the exception to the process was the split between the doors. In order to ensure proper alignment, the doors were fixed in a closed position using tape on the backside.

Pictured here are all of the items needed to weather the underside of the model. This is the same process used on all of the models within this book. The items needed are: a mixing palette, Tamiya putty, Tamiya extra thin cement, Tamiya acrylic Flat Earth XF-52, Winsor & Newton oil paints, Turpenoid thinner (for the oils) and finally Verlinden Productions static grass.

To begin, a dollop of Tamiya putty is placed in a palette well and static grass is placed in the adjacent well. Then Tamiya extra thin cement is added to the putty (as a thinner) and mixed to the consistency of thick paint. Several dollops of Tamiya acrylic Flat Earth are then added to the messy concoction.

Using an old stiff brush, a bit of the concoction is picked up by the brush and then dipped into the static grass.

Weathering

When the markings were completely dry, I washed the entire model with black Winsor & Newton oil paint heavily thinned with Turpenoid. I slightly accelerated the drying time by using a hair dryer. When the Turpenoid from the initial wash had evaporated, I detail washed all of the bolts and other raised details in an effort to make these features stand out.

When the detail wash was dry, I dry-brushed the entire model using Winsor & Newton Yellow Ochre and Olive Green. Each of these colors was dry-brushed over the corresponding color previously painted on the vehicle.

The final stage was to spray the entire model with Testors' Dullcote from an airbrush and highlight all of the nut and bolt detail/seams/join lines with black pastel. I also added paint chips and worn areas with Prismacolor pencils. I generally use steel and dark gray and occasionally medium red.

Applying mud

This is the same process used for the application of mud on all of the models within this book. Essentially, I was interested in creating the look of a vehicle that had been through soft terrain and caked with mud and grass. After road use, the caked mud and grass dries and gradually falls off of the wheels leaving a somewhat

Then the recipe is applied to the lower portions of the model with a stippling motion of the brush. Fine sand can also be added to the mixture for additional effect.

This photo shows how the mixture looks during the application process. It is important to note that one should plan the application of this material due to its unforgiving nature. It is helpful to study historical photos or even modern construction equipment to see how dirt and mud stick to tracked vehicles, prior to your attempted application.

Once the texture is dry Winsor & Newton oil paints are added to achieve depth to the applied texture. AA or AAA permanence oils are used due to their fine pigment grain and "non-clumping" properties. For this process, I primarily use Ivory Black and Burnt Sienna mixed with Turpenoid in a sample cup to the consistency of milk.

Next, this mixture of oils and Turpenoid are evenly applied in a wash over all of the previously applied grassy texture.

Chipping is a weathering technique that has developed a following over the past several years. In order to achieve a natural look, one must employ a careful strategy considering everything from the layers of paint to the areas of the vehicle that would have been scraped, scratched, shot or otherwise damaged. Borrowing from my friend and master modeler Marcus Nicholls, a subtle (somewhat random) scuffing and scratching, using Humbrol or Tamiya enamels applied with a 10/0 or 000 brush, tends to look most realistic.

dusty appearance. In order to achieve this look, one needs the following items: a mixing palette, Tamiya putty, Tamiya extra thin cement, Tamiya acrylics Flat Earth XF-52, Winsor & Newton oil paints, Turpenoid thinner (for the oils) and finally Verlinden Productions static grass. To begin, a dollop of Tamiya putty is placed in a palette well and static grass is placed in the adjacent well. Then Tamiya extra thin cement is added to the putty (as a thinner) and mixed to the consistency of thick paint. Several dollops of Tamiya acrylics Flat Earth are then added to the mixture. Using an old stiff brush, a bit of the concoction is picked up by the brush and then dipped into the static grass.

The radio was wired after the fact (which I do not recommend). As a result, portions of the interior needed to be re-done. Luckily, I hadn't weathered the interior at this stage. The open connection at the upper left side of the radio is for the headphones. The headphones for the figure were later plugged in here.

A view of the interior of the right-hand side reveals the painted details such as the 7.5cm ready rounds and the rolled tarpaulin. The ready rounds were sprayed using a metallic brass enamel that I bought at a craft shop. The tops of the rounds were then hand painted with black and the fuses were painted silver. Tamiya Clear Smoke was then sprayed around the base to provide a weathered look. The tarpaulin was painted using Tamiya and Vallejo acrylics.

It is important to note that one should carefully plan the application of this material due to its unforgiving nature. It is helpful to study historical photos or even modern construction equipment to see how dirt and mud stick to tracked vehicles, prior to your attempted application. After deciding on a strategy, the mud mixture was applied to the lower portions of the model with a stippling motion of the brush. Fine sand can also be added to the mixture for additional effect.

Once the "mud and debris" was dry, a wash of Winsor & Newton oil paints Ivory Black and Burnt Sienna were mixed with Turpenoid in a sample cup to the consistency of milk and applied to the "mud" to achieve depth.

Chipping

Chipping is a weathering technique that has been around for years, but has only recently developed a sizeable following. There are two basic ways to approach this effect. One is achieved by simply painting "chips" over the top of the basic paint scheme. The other is more of a layered effect, which is accomplished by painting one layer, allowing it to dry, then randomly apply liquid mask with a piece of open cell foam on the surface of the model where the crew would have worn the paint. After the mask is dried another layer of paint is sprayed over the mask and allowed to dry. Finally, when the mask is removed, the appearance is that of chipped paint.

Whatever the method chosen, chipping is a technique that when employed correctly can be stunning in its realism. On the other hand, if it is executed poorly, the model can appear contrived and toy-like. In order to achieve a natural look, one must employ a subtle, careful strategy considering everything from setting (as in a vignette or diorama), climate, season, weather, time in service, and other operational conditions such as frequency of maintenance, terrain, etc.

Over the years, I've been impressed with the subtle application of this technique by my friend and master modeler Marcus Nicholls. A review of his work reveals a subtle somewhat random scuffing and scratching. He uses Vallejo acrylics applied with a 10/0 or 000 brush strategically placed on the model. He seems to have a knack placing the right amount of paint in the right location to produce a very convincing result.

For the models in this book, I have employed a similar method to that of Mr Nicholls. While I am a huge fan of Vallejo acrylics, I used Tamiya enamels by virtue of their exact color match to the Tamiya acrylics I used for the base paint scheme.

The final resting place for the superb Plus Models bucket. A perusal of historical photos reveals that a bucket hanging on the exterior of a vehicle was quite a common practice. A couple of dents were added for realism.

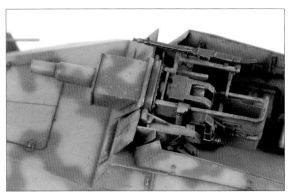

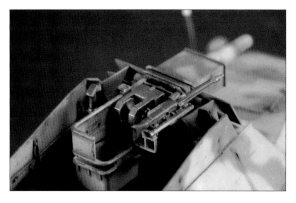

ABOVE AND ABOVE RIGHT Two views of the KwK 37 (L/24) gun. The gun itself, as well as the MG42, was painted by laying down a base coat of Tamiya XF-1 Black and then dry-brushing a mixture of Burnt Sienna, Black and Titanium White Winsor &

Newton Oils (mixed to the color of concrete) and S&J Spray Metal polishing powder over the surface. This mixture is then buffed with black dry pastel. A blued effect can be applied using Tamiya Clear Blue.

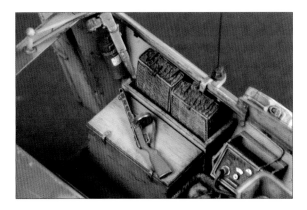

Another view of the interior displaying wicker ammo containers and the more prominently displayed Russian PPSh sub-machine gun taken from the Tamiya Russian assault infantry set (item 35207).

One of the more impressive accessories is the water can by Bego. The cross was made by masking with Tamiya tape. The chipping effect was done as described previously.

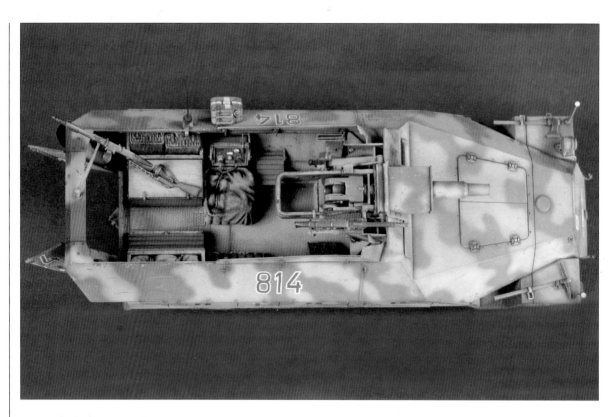

PAGES 44–45 **Multiple views of the finished model.**

SdKfz 251/7 engineer vehicle with assault bridge

Subject:	SdKfz 251/7 engineer vehicle with assault bridge
Model by:	Robert Oehler
Skill level:	Master
Basic kit:	Tamiya SdKfz 251/1 D (item 35195)
Additional detailing sets:	Royal Model RM-086 SdKfz 251 D photo-etch and resin, RM-240 SdKfz 251 Ausf. D details (part 2), R&J Enterprises SdKfz 251 combo set-engine and compartment (RJ35256), Verlinden Productions (0739) assault bridge, Verlinden Productions (0564) SdKfz 251/9 Ausf. D update set, Jordi Rubio (AC-01) German mudguard clearance poles, Evergreen sheet plastic, Modelkasten SdKfz 251 tracks, AFV Club (35007) wheels and tracks.
Markings:	Archer fine transfers, Waffen-SS Panzer/Panzer Grenadier division markings, Eduard German numbers medium (XT-008), Eduard German crosses small late war (XT-002)
Paints:	Tamiya XF-60 Dark Yellow, XF-64 Red Brown, XF-58 Olive Green, XF-57 Buff, XF-10 Brown, XF-52 Flat Earth, Tamiya Enamels XF-60 Dark Yellow, XF-9 Hull Red, XF-1 Black, Testors' Dullcote Laquer-Flat

I have always been interested in the look of the engineer variant. As a result, it was at the top of my list for superdetailing as one of the features of this book. I had decided that the Tamiya SdKfz 251/1D would be the base for this effort and I originally thought I might use the MR Models detail set for this variant. However,

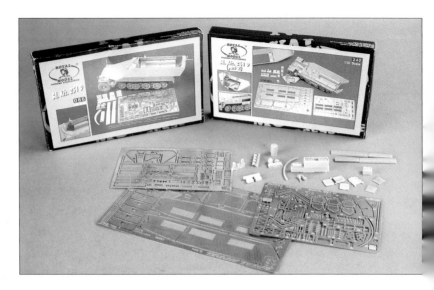

Royal Model photo-etch and resin "goodies" used to enhance the Tamiya SdKfz 251/1 D (Item numbers 086 and 240).

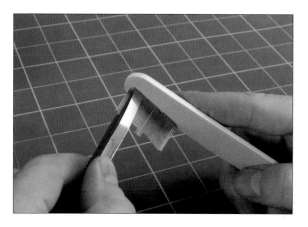

Construction of the model started with the assembly of the many photo-etch details. Each detail was cut off its carrier sheet and each edge was lightly sanded to ensure a straight, clean edge.

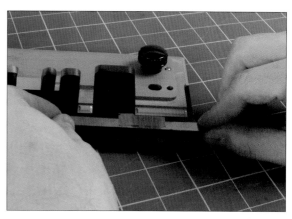

The 8in. Hold and Fold (kindly provided by the Small Shop, EU) was used as a bending brake. To complete a perfect bend, place the photo-etch under the edge of the brake, tighten the holding knobs and then, using the flat blade, lift upward until the desired bend angle is achieved.

after a quick look at the parts, I decided that they weren't up to standard. After a search through the available conversion sets, I decided that the best approach was to use bits and pieces from several different manufacturers detail sets and fabricate what I couldn't find.

The first elements I chose were from Royal Model (kindly supplied by Bill Miley of Chesapeake Model Designs). If you are not familiar with this line of products, I can only state that Royal Model produces some of the most exacting aftermarket products available. The new suspension items from AFV Club, Modelkasten tracks, an engine from R&J Enterprises and the engine compartment as well as the assault bridge from Verlinden Productions rounded out the list of supplies. Thanks to the folks at both R&J Enterprises and Verlinden Productions for supplying these most necessary items.

Construction of the model started with the assembly of the many photo-etch details. Each detail was cut off its carrier sheet and each edge was lightly sanded to ensure a straight, clean line.

The result is a very cleanly formed photo-etch detail without any additional bends or anomalies in the surface of the photo-etch.

My soldering set-up is very basic. It includes: A Weller adjustable soldering iron with cleaning sponge, flat pliers, paste flux, tinning fluid, a flat brush, tweezers, micro shears and of course, solder (not shown).

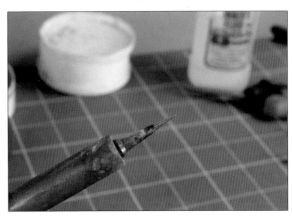

Step one: Ensure the tip of the soldering iron is clean and the iron is pre-heated.

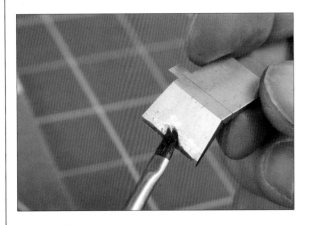

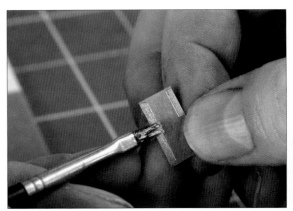

ABOVE AND ABOVE RIGHT **Step Two:** Place paste flux onto both surfaces to be joined.

The 8in. Hold and Fold (kindly provided by the Small Shop, EU) as well as the Etch Mate (from Mission Models) were used to form the Royal Model photo-etch. In my experience I found both bending tools to be well made and easy to use. I also noted that the Hold and Fold, due to its dual knob design, worked well for smaller parts. In general I used the Etch Mate for the larger items and the Hold and Fold for the very tiny parts. Whether you choose one or both, the result is a very cleanly formed photo-etch detail without any unwanted dents or bends in the surface. Once all of the photo-etch bending was complete, I moved onto the soldering.

Soldering

Soldering is the preferred technique used to join metal objects together. While one can certainly use cyanoacrylate glue or two-part epoxy to join metal parts, neither option offers the strength and flexibility of solder.

If you've never attempted soldering, it can be a little tricky at first. I've developed my technique over several years of trial and error (and more than a few curses). The setup requires only basic materials such as a soldering iron with cleaning sponge, flat pliers, paste flux, tinning fluid, a flat brush, tweezers, micro shears and, of course, solder. I found that an adjustable temperature

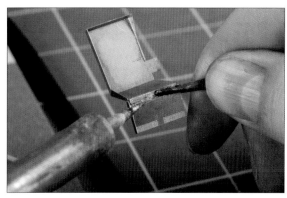

Step Three: Dip the tip of the hot iron into a bead of tinning fluid and then melt a small amount of solder onto the tip of the iron.
Step Four: Apply the tip of the soldering iron to the area to be joined. The hot solder will make contact with the flux and, with a little assistance, will cover the area to be joined. This process is known as "tinning."

Step Five: Once both sides of the join areas are tinned, I apply another thin layer of flux to each previously tinned area. Then I position the parts and hold them in place using the tweezers. When the hot iron is then applied to the join area, it melts the solder and it flows into the entire gap fixing the parts together. It is important to note that positioning should be checked to that parts are not inadvertently soldered misaligned. Also, one should take care not to solder the tweezers to the parts.

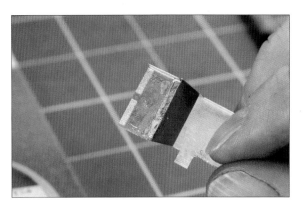
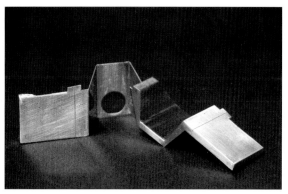

Voila! A perfectly soldered sub-assembly.

Soldered and polished fender sections. After soldering, I clean/wipe off the flux paste and use 1,200-grit sandpaper and a burnisher to lightly polish the brass. This process removes surface anomalies and provides ready surface for priming.

soldering iron offers the most flexibility to the modeler due to its ability to handle solder of differing temperatures.

Running gear

All of the suspension components for the model were replaced with the AFV Club wheels and tracks set. This effort required some minor surgery and a good deal of test-fitting. It is, however, worth the effort considering the gain in accuracy, detail and the ability to articulate the suspension (an option I did not choose).

Storage racks

One of the items that helped me make the decision not to use the MR Models detail set was the storage racks. In the MR Models set, all of the items stowed in these racks were molded together in one solid lump of resin. Disappointed, it became apparent that if I wanted racks that would allow the stowage of individual detail and crew equipment, I needed to fabricate them myself. The racks that I am referring to are located on top of the hinged storage boxes at the

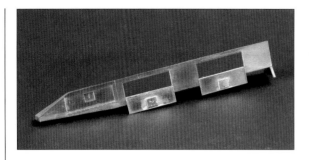
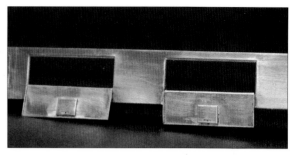

ABOVE AND ABOVE RIGHT Two views of the sponson lockers after assembly, soldering and polishing. The lock box on the center of each hinged door, and the internal frame, were soldered in place separately. As you can see, proper alignment is extremely important here.

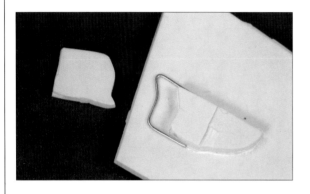
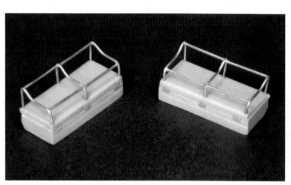

In order to create the loose storage racks located on top of the rear hinged boxes, a jig was created, using an old piece of discarded resin sheet, to consistently bend the brass wire. To use the jig, the brass wire was bent around the form and the free piece of the jig was pushed onto the top of the fixed portion of the jig to achieve the final form.

With the end and middle sections of the bin formed, additional lengths of brass rod were soldered to the previously bent sections as shown. The racks were then fixed into pre-drilled holes using CA glue.

back of the fighting compartment. After careful measuring was done, I decided to use two of the front resin boxes from the MR Models set. These were cut to size and were carefully sanded and scribed to once again resemble storage bins. Eduard hinges were used on each of the bins for an additional touch of realism.

To create the upper portion of the rack, a jig was created to consistently bend the brass wire. Using two pieces of discarded resin sheet, I carved two small pieces of the resin into the shape of the bracket and glued the lower piece to a larger resin sheet with cyanoacrylate glue. To use the jig, the brass wire was bent around the form and the free piece of the jig was pushed onto the top of the fixed portion of the jig to achieve the final form. I formed three of these pieces for each of the stowage boxes for a total of six.

With the end and middle sections of the bin formed, additional lengths of brass rod were cut and soldered to the previously bent sections as described in the photo captions. The racks were then placed into pre-drilled holes and set using cyanoacrylate glue. As a final step, individual pieces of stowage were test-fitted to these bins. These simulated items included pioneer sacs, explosive charges, Teller mines, wire reels etc.

The engine

Another simple idea that sadly went awry was the engine. It was my original intention to find a simple drop-in set for the engine and engine compartment. Unfortunately, it was not to be.

LEFT AND BELOW LEFT Left and right views of the engine with its top off. The engine was a carefully adjusted and fitted amalgamation of several aftermarket manufacturer's offerings. It consists of the R&J Enterprises engine block, carburetor and radiator; Verlinden Productions air cleaner, fluid reservoir brass and fire extinguishers. And finally the white styrene items: belt pulleys, manifold, etc. were formed from Evergreen plastic

A view of the engine with its top on. As you can see, it is a rather cramped and busy endeavor. Also visible are the engine locking latches in Evergreen white styrene.

The underside of the engine compartment displaying the carefully cut out underside of the Tamiya kit with the carefully fitted AFV Club axel and R&J Enterprises engine. Note the opening between the drives that was later closed on the finished model.

 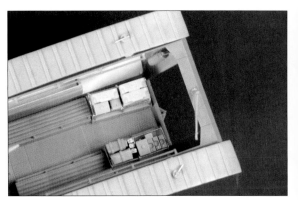

The roof section of the fighting compartment displaying the Royal Model photo-etched hinges and head pads.

A close-up view of the fighting compartment displaying the MR Models charges, Verlinden Productions engineer bags and the Plus Model Teller mine cases.

Once again, after a review of my references, I was surprised by the limited offerings. Just by chance I mentioned my interest in finding an engine to include as part of this book to Rich Sullivan at R&J Enterprises. To my surprise, he had recently received a master of a SdKfz 251 engine and was just about to get the first samples in resin. It was good of him to send me one of those samples straight away, because it served as the basis for my engine compartment. The R&J set includes a portion of the chassis, the block, radiator and all of the other important components. While the scale was spot on, some of the components were warped or had unseen bubbles (discovered later) in unfortunate locations. Also, fitting the AFV axle and springs for the front wheels together with the R&J items required a little surgery. In any case, the gray resin you see in the photos are the R&J Enterprises components, the green resin and photo-etch are from Verlinden Productions and the white styrene components were my contribution thanks to Plastruct and Evergreen.

The constant measuring, trimming, adjusting and test-fitting seemed endless. However, in the end I was able to create a reasonable facsimile of the Maybach HL 42 six-cylinder water-cooled in-line engine and its surrounding compartment from the amalgamation of parts from the different manufacturers. By the way, the belts are strips of Tamiya tape.

Interior details

With the exception of the storage bins at the rear of the fighting compartment, the interior is visually similar to the standard SdKfz 251/1 D. Many of the same basic improvements were made to this vehicle that were made on the previously described "Stummel," (items such as the side armor and bolt detail, the addition of the clutch brake and throttle pedals, and finally the wiring of the radio). In addition to the stowage improvements, I added a great deal of small detail to add interest and provide the viewer a glimpse into how the combat engineers of the German military lived. In order to accomplish their mission, engineers carried a lot of gear and specialist equipment—things like explosive charges, mines, spools of wire, lane-marking flags and other equipment for creating and marking minefields, etc. Similarly, I have added a great deal of equipment and crew items to the model to give the impression that the vehicle was "lived in." A close look in the fighting compartment reveals the MR Models 5 and 10kg charges, Verlinden Productions engineer bags, the Plus Model Teller mine cases and water containers, wire spools, Tamiya panzerfausts and, finally, a grenade case.

The forward armored visors and side vision blocks were also added to increase the realism of the piece. The forward visors were almost always open in reference photos, so I thought it would be interesting and include them open on the model.

Royal Model provides superb resin visors as well as extremely well engineered photo-etch for the interior hinge mechanisms. The side vision blocks were simply resin additions, which required no further photo-etch enhancements.

Rounding out the description of the added interior details, I also made improvements to the rear crew door. The latch, chain and lock on the upper portion of the crew door came from Royal Model. These items simply replaced the molded-on detail present in the kit. An additional touch was the tiny springs that were taken from my box of tiny microelectronic items initially provided by my friend Stan Spooner.

Exterior details

Between the Royal Model photo-etch, the AFV Club suspension improvements, Modelkasten tracks and the addition of the Verlinden Productions assault bridge (item 0739); the base kit has become a new vision of its former self.

In addition to the previously described engine and suspension details, other obvious enhancements include the brass fender and side storage lockers, improved detail around the muffler and the assault-bridge brackets.

A close-up view of the side storage areas reveals the MR Models brackets and the wooden planks, which are actually stirring sticks liberated from the local Starbucks coffee shop. I was fortunate to find these stirring sticks due to their realistic scale, texture and grain. In order to make them useable, they were first coated with thin cyanoacrylate glue and then secured with the same glue in a side-by-side orientation. When dry, they were carefully cut and fitted to the pre-positioned brackets.

Due to the previously described soldering process, the fenders and open sponson storage lockers were very strong and resilient. Because I knew these items would be handled a great deal during the construction and painting phases, I attached them to the hull exterior with two-part epoxy for additional strength. I placed items representing the personal effects of the crew in the two storage locker compartments. These items are intended to encourage the imagination of the casual observer to appreciate the conditions under which the crew lived.

The small "T" handle on the outside of the crew door was fashioned from Plastruct styrene rod and glued into position.

Putting it all together ... a front three-quarter view of the model reveals many of the added enhancements to the Tamiya base kit. One can clearly see the open visors, the Verlinden Productions assault bridge, Modelkasten track, Jordi Rubio width indicators, AFV Club front and road wheels and all of the other details.

Overall top view of the model prior to painting.

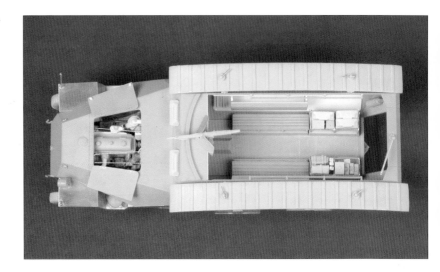

Ground level front view of the model prior to painting.

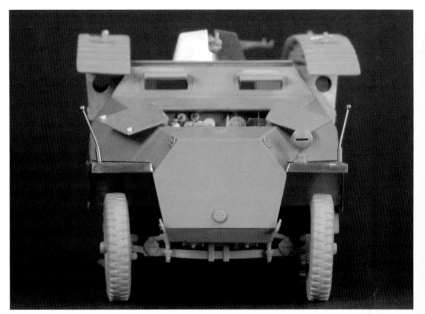

Three-quarter view of the entire model, with fitted sub-assemblies, prior to painting.

Close-up view of the side storage areas revealing the MR Models brackets and the wooden stirring sticks liberated from the local Starbucks coffee shop. I really like the scale texture and grain of the sticks. They were first coated with thin CA glue and then affixed side-by-side. When dry, they were carefully cut and fitted to the pre-positioned brackets.

Detail of the lower chassis showing the brass fender, drive sprocket and muffler.

The weapons typically assigned to the SdKfz 251/7 D were two MG42 machine guns carried both forward and aft. These were designed as primary defensive weapons. Both MG42s featured on the model came from one of the Dragon figure sets and the ammo belt was taken from the accessories featured in the Tamiya 251 base kit.

Painting and weathering

The model was built in sub-assemblies to later accommodate the painting effort. Plasticote automotive primer was used to prime all sub-components of the model and, due to its color, was also used as a base coat for the rest of the painting process.

Similar to the method I used on the previous models, I sprayed Tamiya acrylics Dark Yellow, XF-60, across the entire model. Then the highlighted look was achieved by carefully highlighting, spraying the inner portions of each panel section with Tamiya acrylics Dark Yellow, XF-60, leaving a darker yellow/brown edge on the outside of each. When the highlighting was satisfactory on both the interior and exterior, I began painting the camouflage. Since this was a late-war vehicle with a heavily mottled three-color scheme (patterned after the photograph on page 35 of Osprey's New Vanguard on the

The upper hull is painted with XF-60 Dark Yellow, by carefully highlighting the inside of each panel. Lower hull showing primer undercoat.

Close-up of the model with all of the highlighting completed.

ABOVE AND ABOVE RIGHT Two views of the freshly Dullcote-sealed camouflage with all of the sub-assemblies in place. At this stage, even without the wheels and tracks, the model is beginning to really take shape.

After all the highlighting and camouflage was completed, the entire model was, once again, broken down into sub assemblies in order to first apply Testors' Dullcote from the rattle can and then to apply the washes/weathering to the interior components.

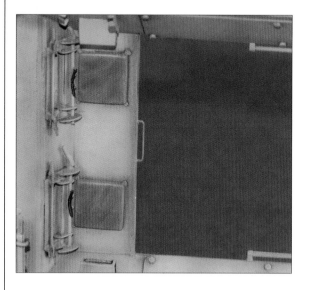

View of the hull roof above the driver and radio operator. The most prominent features are the head pads rendered in brown leather Vallejo acrylics. Of note are the opening visor assemblies looking toward the front to the vehicle as well as the fixed vision blocks on the left and right sides of the vehicle.

SdKfz 251), I was interested in a dark look with a great deal of contrasting highlights. To accomplish this look I began laying in the camouflage with Tamiya Red Brown, XF-64, using what I have come to call the "scribble method." I refer to it as the "scribble method" because it closely resembles a random scribble across the surface of the model. The initial scribbles are applied in a small area with very thin paint (40:60 paint to thinner) at a low pressure (about 12–15psi). Once I am satisfied with the look, I add more paint definition to the areas where the scribbles cross. It is important to carefully watch your efforts when embellishing the crossing areas to ensure that the overall look remains random. I then lay in the olive green, in the same fashion, adjacent to the previously applied red brown pattern.

Next, I detail-washed the entire model with Black Winsor & Newton oil paint heavily thinned with Turpenoid to pick out the recessed and raised features. Once again, this wash was applied by placing the mixture with a liner brush next to (or onto) the detail to be enhanced. Capillary action takes the paint around the detail without any additional assistance. Once the Turpenoid had evaporated, I dry-brushed the entire model using Winsor & Newton Yellow Ochre, Burnt Sienna, and Olive Green. Each of these colors was lightly dry-brushed over the corresponding color, previously airbrushed onto the vehicle. After drying for several hours, the model was sprayed with Testors' Dullcote.

At this point the upper hull and the bulkhead between the engine compartment and the driver were still not fixed in place. Therefore, painting the engine compartment was relatively easy. Each component was carefully blocked in using a brush and Vallejo acrylics. A detail wash was applied to each of the components along with a subtle bit of chipping.

The engine was initially painted black and then was highlighted with a very thin mixture of neutral gray and brown using the Custom Micron-SB. Afterwards, the engine was carefully treated (dry-brushed) with graphite and then was given the black pastel treatment in all of the recesses using a small pointed brush.

The front license plate is a waterslide decal taken directly from the Tamiya kit. In order to ensure a snug seating of the decal, I spray the areas where the decal will be applied with Future floor wax.

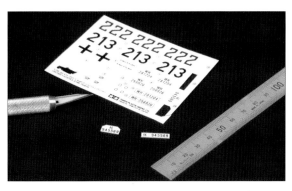

When using waterslide decals it is important to carefully trim off the carrier film from around the decal. Trimming is most easily accomplished with a new blade, a straightedge and a soft cutting surface.

The vehicle I chose to render is from the 12th SS "Hitler Jugend" Division. Vehicles in this division carried their divisional insignia near the starboard visor (as shown) and another, larger, insignia on the rear starboard panel. Archer Fine Transfers offer a range of sizes of this divisional insignia within their extensive range of dry transfers. As you can see, this one fit perfectly and was a cinch to use.

Modelkasten tracks (Item K-019) are finely detailed, easy to assemble and have a realistic appearance when painted and weathered.

Markings

In the course of determining the paint scheme for the model, I found a photo on page 35 of Osprey's New Vanguard on the SdKfz 251. The vehicle represented in the photo was number 440 from the 12th SS "Hitler Jugend" Division. In order to replicate the markings from the photo, I used the excellent Eduard German numbers medium stencil for the numerical markings. The Eduard vinyl stencils are simple to use. In this case, I was only spraying black numerals so it was simply a matter of placing the stencil and spraying.

Another unique thing about this vehicle was the divisional insignia carried near the starboard visor and another, larger, insignia on the rear starboard panel. Archer Fine Transfers offers a range of sizes of this divisional insignia within their extensive offerings of dry transfers. As you can see by the photo, this one fit perfectly.

The front license plate was a waterslide decal taken directly from the Tamiya kit. When using waterslide decals it is important to carefully trim off the carrier film from around the decal.

When placing the decal onto a flat paint surface it is advisable to ensure a snug seating of the decal with decal setting solution. Laying the decal onto an area previously sprayed with Future acrylic floor wax can also enhance the seating of the decal. See the SdKfz 251/1 C portion of the book for the application sequence.

Modelers have to be creative to achieve their goals. One example of this creative utility is shown here. In order to uniformly paint and weather the tracks, I simply find a large roll of 3in. masking tape, reverse a length so the adhesive is face out and then lay on the tracks. The tracks are held in place firmly, but the tape will not pull off the freshly applied paint. I complete each step of the finishing process, from base coat to pastels on this roll.

In an effort to simulate the wear of the road wheels on the dusty surface of the track, I "paint" a line of black pastel mixed with Turpenoid on both sides of the guide horn. Then, using a thin piece of Tamiya tape, I mask the area and apply the weathering pastels over the top. When they pastels are dry, I remove the tape and gently rub the previously masked, black pastel with graphite on a soft, stiff brush.

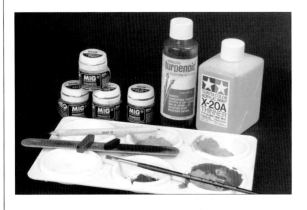 

When weathering tracks, I usually use pastels, both wet and dry. Shown here are the tools I use to mix and apply the pastels. Mig pigments (or Migments as I call them) are mixed in the palette with either the Tamiya acrylic thinner or Turpenoid and are applied to the tracks/model with a stiff brush. MMP Powders (not shown) are also used interchangeably in this process.

A clear view of the finished tracks prior to the graphite treatment (top) and the black pastel line on both sides of the guide horn prior to applying the weathering pastels.

ABOVE AND ABOVE RIGHT Two views of the finished spare tracks on the finished model. Note the realistic wear of the road wheels against and on either side of each guide horn. The graphite over the pastel really makes the difference.

Tracks and tires

In order to uniformly paint and weather the tracks, I simply found a large roll of 3in. masking tape, reversed a length so the adhesive was face out and then laid on the tracks. The tracks were held in place firmly, and the freshly applied paint was not pulled off. I completed each step of the finishing process, from base coat to pastels on this roll. The tracks were then easily removed and reversed to facilitate the painting on the other side.

In an effort to simulate the wear of the road wheels on the dusty surface of the track, I "painted" a line of black pastel mixed with Turpenoid on both sides of the guide horn. Then, using a thin piece of Tamiya tape, I masked the area and applied the weathering pastels over the top. When the pastels were dry, I removed the tape and gently rubbed the previously masked black pastel with graphite on a soft, stiff brush.

The beauty of pastels as a weathering medium is that they can be used both wet and dry. In weathering the tracks, the initial effort was wet. Mig pigments were mixed in the palette with either the Tamiya acrylic thinner or Turpenoid and were generously applied to the tracks/model with a stiff brush. MMP Powders were also used interchangeably in this process.

I painted wheels by initially spray-painting the hub section. Then I blocked-in the tire using a black/gray/brown mixture of Tamiya enamels applied by brush. Finally, I used an airbrush to apply an even layer of Tamiya acrylic Flat Earth, XF-52, around the hub as well as on and around the tread. The final step was to apply a black/brown/gray mixture of dry pastel to the tread and sidewall of the tire gently rubbed with my finger. Touch-ups were completed with pastel applied with a brush.

I will generally paint wheels by painting the hub section first. Then I will block in the tire with a paintbrush using a black/gray/brown mixture of Tamiya enamels. Finally, as shown, I use an airbrush to apply an even layer of Tamiya acrylic XF-52 Flat Earth, around the hub as well as on and around the tread. The final step is to apply a black/brown/gray mixture of dry pastel to the tread and sidewall of the tire gently rubbed with my finger. Touch-ups are completed with pastel applied with a brush.

Joining the upper and lower hull

With the interior painting complete the upper and lower hull can be joined. I, once again, used Tamiya extra thin liquid cement as mentioned in the previous chapter.

Weathering the underside of the model

See the section on applying mud in the chapter on the SdKfz 251/9 "Stummel."

Chipping

The chipping was applied to the model as previously described using Tamiya enamels and a 10/0 brush. I strategically placed random scuffs and scratched on portions of the vehicle that would have seen this type of wear. On the dark yellow I placed hull red (primer) scuffs and scratches and on the red brown and olive green, I placed dark yellow scratches. To add further visual interest, I occasionally placed hull red on the edges of the dark yellow scratches to give the impression that the paint was scratched down to the primer.

ABOVE AND ABOVE RIGHT Two different views of the weathering on the underside of the vehicle created using putty, paint (both acrylic and oil), static grass and pastel. For the description of this weathering process see the SdKfz 251/9 section of the book.

Soldier's gear in scale. These are some of the details that give the model a more interesting, lived-in look. Shown here are many of the small details that will adorn the interior of the model.

Close up of the grenade case and the crew's personal gear hanging from the grab handle. Note the way the paint has been scratched away as in real life. The metallic paint scratches and chipping was achieved using Berol Prismacolor pencils.

Right side view of the fighting compartment displaying the crew stowage.

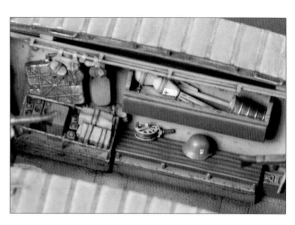

Left side view of the fighting compartment. Note the mixture of personal items with weapons and sustenance items. It is a functional clutter seen in combat conveyances throughout the ages.

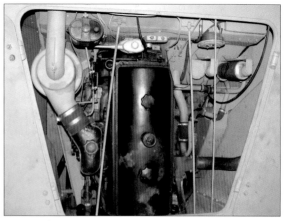

ABOVE AND ABOVE RIGHT My rendering of the engine compartment and the almost accurately restored "real thing." As you can see, all of the major components are present to form a convincingly dirty facsimile similar to that of the real thing.

An overhead view of the business end of the vehicle showing the open engine compartment. This kind of detail offers an additional level of interest.

The open sponson storage lockers. Open compartments featuring the personal effects of the crew spark the observer's imagination offering a glimpse into the conditions in which the crew lived.

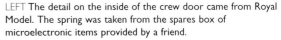

The small "T" handle on the outside of the crew door was fashioned from Plastruct styrene rod and glued into position.

LEFT The detail on the inside of the crew door came from Royal Model. The spring was taken from the spares box of microelectronic items provided by a friend.

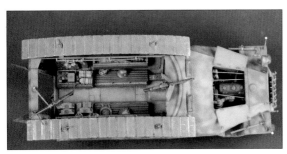

Overhead view of the entire vehicle provides a perspective on how busy and cramped life must have been for the seven crewmembers.

The completed Plus Models (item 152) buckets sitting on a dime for perspective. These little gems are easy to make and add a touch of realism to any subject.

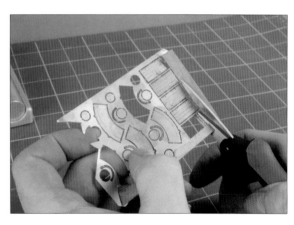

To make the buckets, the first step is to cut the photo-etch items off of the fret. One should also inspect the photo-etch for surface burrs etc. A light sanding on the edges should take care of any issues.

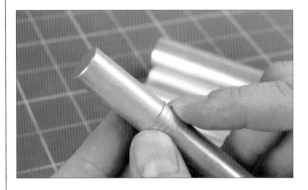

I used the Brass Assist tool (once again, kindly provided by the Small Shop, EU) for this project. When bending brass, It is best to begin with the largest roller and work your way down to the size just a little smaller than the diameter of the actual bucket. Due to the softness of the Plus Model's brass, no annealing was necessary.

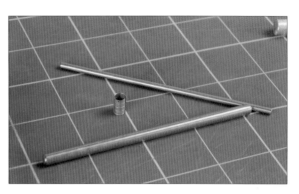

Several diameters of rod were used to bend the brass into the final size of the actual bucket as seen here.

In order to achieve the final shape of the bucket, I simply pressed the semi-formed brass around the rod until the final diameter was achieved. It is also important to ensure that the both ends of the seam meet by themselves. This positioning consideration is important when the heat is applied and the joint is soldered.

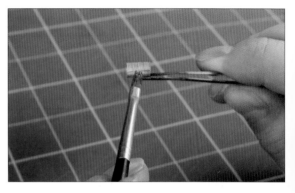

The next step is to brush flux paste onto both sides of the seam. Then apply solder to each portion of the seam, effectively "tinning" the areas to be joined as described earlier in the chapter.

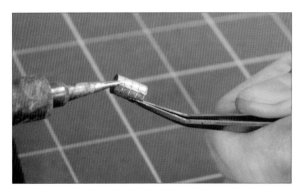 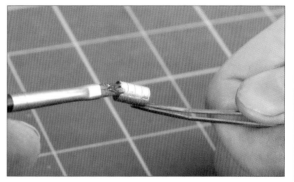

Then, carefully align the seam, hold it lightly using a tweezers or a heat sink and apply heat. As mentioned earlier, the better the alignment at the seam, the less positioning and holding is required. When the heat is removed and the seam joint is firm, you should have a nicely soldered cylinder.

The final step is soldering the bottom of the bucket into place. Since the bottom is only a small disk, all of the earlier alignment effort at the seam will pay off. If the seam alignment is correct, the bottom (disk) should drop through the top of the cylinder and seat at the bottom. I will generally apply flux paste to the bottom, and tin the inside of the cylinder and then apply flux paste to the bottom inside prior to dropping the disk. Once the disk is in place, all that must be done is to apply heat. Important note: The tip of the iron should be applied to the bottom of the cylinder furthest away from the seam to keep it from readjusting or opening altogether.

LEFT With the soldering complete the form is now that of a scale bucket.

BELOW Left side view of the finished model.

Right side view of the finished model

This view favors the divisional marking position unique to the 12th SS "Hitler Jugend" (kindly supplied by Archer Fine Transfers).

The weapons typically assigned to the SdKfz 251/7 D of this period were two MG42s carried both forward and aft. These were designed as primary defensive weapons. Both MG42s featured on the model are from one of the Dragon figure sets and the ammo belt was taken from the accessories

ABOVE AND BELOW **Left and right side views of the finished model.**

A gracious gift

Subject:	*"A Fine Day Out" Diorama*
Modeler:	*Robert Oehler*
Skill Level:	*Intermediate*

While I was trying to decide on a diorama setting for this feature, I happened to be visiting St. Louis for the VLS MasterCon event. While I was there Charlie Pritchett and I decided to pay Mr François Verlinden a visit at Verlinden Productions studios. While I wandered through the displays of dioramas that have so profoundly influenced my modeling, I happened upon and immediately fell in love with a base that François was preparing for painting. In the course of conveying my compliments, François offered to let me have it for use in the book. I gratefully accepted the offer! As you can see by the photos, the base conveys a quiet little roadside shrine in the shade of a tree. The base itself measures 13 x 13in. (32 x 32cm) and included the shrine and the tree.

Composing the diorama

Now that I had a satisfactory base, I needed to create a story involving one of the 251s, some figures and perhaps another vehicle. Since the base was so pastoral, I wanted a simple and relaxed scene. After an extensive search for the appropriate figures, I settled on the idea of converting several Verlinden Productions figures (from set number 1916) and the radio operator from the self-propelled gun crew, produced by Tristar (item 002). I chose these figures due to their body language. Their bodies convey tired, but confident poses and their facial expressions are stern and purposeful. I used one Warriors head on the officer and a head from another Verlinden Productions figure for the officer's companion.

The figures in front of the shrine are basically Verlinden figures pieced together from several different sets. The head of the officer on the right is from Warriors.

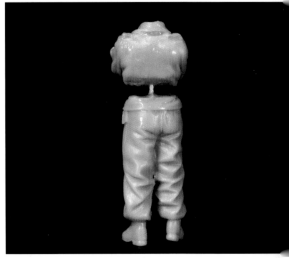

Rear shot of the figure with the folded arms. As you can see, he is pinned together at the waist with a length of paper clip.

ABOVE Close-up of both figures. The headsculpts are nicely done.

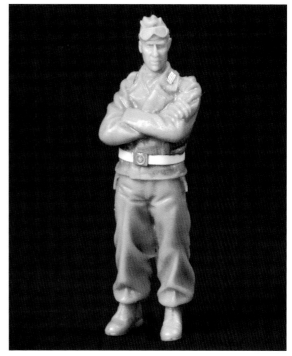

RIGHT The figure with the folded arms prior to painting. His center has been filled/sculpted with Duro Putty using a toothpick. His belt is added using a length of strip plastic and the brass details like the buckle and Waffenfarbe are from Lion Roar.

I also decided that another vehicle was necessary in the foreground. After placing several different candidates onto the base for evaluation, I decided the Kubelwagen with the infantry cart fit the bill nicely. I spent a good deal of time adjusting the overall composition until I found the right positions for each element. This staging exercise, prior to painting, is and important first step to ensure that all of the parts "fit" and contribute to the story the scene tells.

Painting the base

Painting the base began with the cobblestone road and the retaining wall. The base coat for the road was sprayed with a mixture of Tamiya acrylics (four parts XF-57, two parts XF-20 and one part XF-52) using the Iwata Revolution. The retaining wall was painted with a mixture of four parts XF-57, two parts XF-20 and one part XF-2, once again applied with the Revolution.

Overhead shot of completed figures in situ.

Frontal close-up of figures. I chose these figures due to their body language. Their bodies convey tired, but confident poses and their facial expressions are stern and purposeful.

Detail close-up of the officer and his companion. The Waffenfarbe is from Archer Fine Transfers. They are easy to use and look great on the figures.

Staging the overall composition prior to painting is and important first step to ensure that all of the parts "fit," in creating a scene that tells a story.

After the road and retaining wall were painted and dry, I painted the ground surfaces with an even coat of Tamiya Flat Earth XF-52 using the Iwata Revolution airbrush. Several of the rocks, both large and small, were deliberately "picked-out" and painted by brush using Humbrol enamels, varying mixtures of sand and cream, to achieve different tones.

To achieve a natural water look, the streambed was painted Tamiya Flat Earth with a slightly greenish hue at the bottom and a darker brown along the banks for depth. These were mixed by eye, so I do not have an exact color reference.

Next static grass was applied to the groundwork and the center of the cobbled roadway using a mixture of white glue and water mixed to the consistency of milk. This white glue mixture was applied to the groundwork with a small, flat paintbrush and then the Verlinden Productions static grass was sprinkled onto the glue. After sprinkling the static grass onto the glue, I gently blew on the area to

The Diorama base was graciously donated to the project by François Verlinden. The base itself measures 13 x 13in. (32 x 32cm) and included the items pictured as well as the tree, seen in subsequent photos, complete with colored foliage.

Painting the road began with the cobblestones and the retaining wall. The base coat for the road was sprayed with a mixture of Tamiya acrylics (four parts XF-57, two parts XF-20 and one part XF-52) using the Iwata Revolution. The retaining wall was painted with a mixture of four parts XF-57, two parts XF-20 and one part XF-2, once again applied with the Revolution.

With the road and wall painted, the ground surfaces were painted with an even coat of Tamiya Flat Earth XF-52 using the Iwata Revolution airbrush. Several of the rocks, both large and small were deliberately "picked-out" and painted by brush using Humbrol Sand and Cream intermixed to varying tones. The stream bed was painted with a slightly greenish tone with a darker brown along the banks for depth.

Next static grass was applied to the groundwork and roadway using a mixture of white glue and water mixed to the consistency of milk. This white glue mixture was applied to the groundwork with a small, flat paintbrush and then the Verlinden Productions static grass was sprinkled onto the glue.

This detail shot of the static grass application reveals the random balance of grass and bare spots. The intention of this application is to create interest as well as simulate nature. In addition to the static grass, longer grass was "planted" in selected, somewhat random locations such as corners and near rocks to create an unkempt look to the shrine area. Holes were drilled and the longer grass was glued in place using CA glue.

When the overall static grass application was thoroughly dry, an oil wash of Winsor & Newton Burnt Umber and Black (roughly a 50/50 mix) was applied directly onto all of the groundwork and allowed to dry.

Once the base was dry, it was time to begin drybrushing. This was done using Humbrol enamels. Successive layers of green and light green (green + yellow) were slightly intermixed to create a subtle green tone over the dry looking static grass.

After the drybrushing was completely cured, I applied additional "spot" washes of oils (black, burnt sienna and raw umber) to blend the road and groundwork. While these washes dried, I blocked in the colors of the shrine.

remove the excess static grass. I purposefully alternated random patches of static grass and bare spots. The intention of this application was to create interest as well as simulate nature.

In addition to the static grass, longer grass was "planted" in selected, somewhat random locations such as corners of the wall and near rocks to create an unattended look to the shrine area. The planting was achieved by drilling holes in the base and then the longer grass was glued in place using cyano glue.

When the overall static grass application was thoroughly dry, an oil wash of Winsor & Newton Burnt Umber and Black (roughly a 50:50 mix) was applied directly onto all of the groundwork and allowed to dry.

Once the base was dry, I dry-brushed Humbrol enamels in successive layers of green and light green (green + yellow) to create a subtle, natural looking green tone over the dry static grass.

After the dry-brushing completely dried, I applied additional "spot" washes of oils (black, burnt sienna and raw umber) to blend the roads and groundwork. While these washes dried, I blocked-in the colors of the shrine.

This detail shot of the basic groundwork with all of the stages complete.

The final step in completing the base was the addition of "standing water." I used a two-part epoxy product called Solid Water (graciously supplied by my friend Marcus Nicholls of *Tamiya Model Magazine International*). To get the appropriate green-brown mixture I was after, I mixed Tamiya acrylics Green XF-4 and Brown XF-10 along with a touch of Mig Pigments to taste.

Once the color of the epoxy was satisfactory, the first layer of the Solid Water product was laid in using a paintbrush. I essentially painted an even coat of the solid water product onto the base in order coat the "water" area and to define the edges of streambed. Due to its thick consistency, it was necessary to "help" the product to flow around the grass and rock elements.

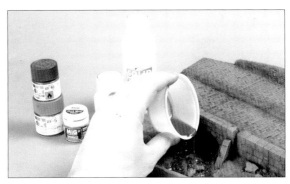

The Solid Water was carefully poured over the top of the previously applied areas making sure that no bubbles were present. When pouring, one should be careful to move around the treatment area to ensure an even application of the product pausing periodically to observe the "water level." It should be noted that if the product is applied too quickly, it will easily overflow its banks, flooding the adjacent areas.

The shrine

After the initial colors of the shrine were blocked-in and were sufficiently dry, I added grout to the brick. Finally, with all of the detail painting completed, a dark wash (50:50 mix of Turpenoid with black and burnt sienna) was applied to tone down the brick and to appropriately weather the item so it blended in to its surroundings on the base. The final step was to dry brush the shrine with the appropriate, corresponding highlight tone to the colors that were originally blocked-in.

Standing water

The final step in completing the base was the addition of "standing water." I used a two-part epoxy product called Solid Water (graciously supplied by my friend Marcus Nicholls of *Tamiya Model Magazine International*). To get the appropriate green-brown mixture I was after, I mixed Tamiya acrylics Green, XF-4, and Brown, XF-10, along with a touch of Mig Pigments to taste.

Once the color of the epoxy was satisfactory, the first layer of the Solid Water product was applied to the streambed using an old paintbrush. This was

ABOVE A close-up view of the completed stream area. When properly colored and carefully applied, the Solid Water epoxy provides a convincing replication of standing water in scale.

RIGHT An overall view of the completed base.

The shrine element with all of the detail painting completed. A dark wash (50:50 mix of Turpenoid with Winsor & Newton Black and Burnt Sienna) was applied after this photo was taken to tone down the brick and to appropriately weather the item so it blended in to its surroundings on the base.

essentially painting an even coat of the solid water product onto the base in order coat the stream bottom and to define the edges of streambed. Due to its thick consistency, it was necessary to "help" the product to flow around the grass and rock elements.

About an hour after the first layer was applied, a second layer of the Solid Water product was carefully poured over the top of the previously applied areas making sure that no bubbles were present. When pouring, one should be careful to move around the treatment area to ensure an even application of the product. In addition, one should pause periodically to observe the "water level." It should be noted that if the product is applied to quickly, it will easily overflow its banks, flooding the adjacent areas.

Shade tree

One of the more interesting features of the diorama is the very convincing Verlinden scale tree (item 1995). Considering that this item consisted of a painted resin trunk and sea foam branches with bits of colored foam glued to them, I was impressed with the realistic look of the completed tree. To "plant" the tree, I simply drilled a hole in the display base as well as the base of the tree. Then I inserted a length of brass rod into the hole at the base of the tree. I then fixed the tree in place using cyanoacrylate glue.

The Kubelwagen and cart

The presentation of the Kubelwagen in the foreground of the diorama consists of the Tamiya kit with Decal Star soft-top sculpted in the folded position. The smiling driver figure is also a heavily converted Tamiya figure with a Hornet head, Verlinden Productions arms and a foil "Y" harness with cartridge pouches.

The infantry cart behind the Kubelwagen is from Dragon Models. Both the cart and the Jerry cans it is carrying are out-of-the-box items with no additional detailing.

Overall, the base was a fun way to spend a couple of afternoons. I was really pleased with the Solid Water product and cannot wait to use it again. While the other elements of the diorama were more involved, I truly enjoyed the process of composing, painting and seeing it all come together.

The very convincing Verlinden Productions scale tree (item 1995). Considering that this item consisted of a painted resin trunk, and sea foam branches with bits of colored foam glued to them, I was impressed with the overall look of the completed tree.

An overview of the entire diorama with all of the elements present.

Assorted shots of the scene.

ABOVE AND BELOW Different views of the Kubelwagen, in studio and in the diorama. The presentation consists of the Tamiya kit with Decal Star soft-top sculpted in the folded position. The smiling driver figure is also a heavily converted Tamiya figure with a Hornet head and foil "Y" harness.

The radio operator figure is a nicely sculpted figure offered by Tristar and the jerry can is from Bego.

ABOVE AND RIGHT The infantry cart (Dragon) is shown prior to painting and in-place on the diorama. Both the cart and the Jerry cans it is carrying are out-of-the-box items with no additional detailing.

RIGHT The photographer trying to frame his subjects for posterity. This convincing figure is basically a Verlinden Productions figure with a right arm and hand from my spare parts bin. The figure is holding a camera in his left hand.

Further reading, media, websites and museums

While references for the SdKfz 251 have been somewhat hard to find, there are a few that I found to be most helpful in the course of writing this books. Each of these titles is oriented to the needs of model builder and historian alike. All of the books mentioned below display photos (some color) and descriptions of the various production models, as well as the mission-oriented variations. Some titles also include scale line drawings and colour reference plates.

Books
Climent, Charles, Armor Series No. 21: *SdKfz 251 in Action* (Squadron/Signal, 1981)
Culver, Bruce, and Feist, Uwe, *Schutzenpanzer* (Ryton Publications, 1996)
Culver, Bruce, Vanguard 32: *The SdKfz 251 Half-Track* (Osprey Publishing, 1983)
Culver, Bruce, New Vanguard 25: *SdKfz 251 Half-Track* (Osprey Military, 1998)
Forty, Jonathan, Military Vehicles in Detail 1: *SdKfz 250/251 Armoured Halftrack* (Ian Allen Publishing, May 2003)
Ledwoch, Janusz, Militaria in detail 1: *SdKfz 251 in Polish Museums* (Wydawnictwo Militaria, 2000)
SdKfz 251, Ground Power Special Issue (Delta Publishing Co. Ltd., February 2002)
Sturm and Drang 3 (Sensha-magazine Co. Ltd., 1991)
Terlisten, Detlev, Nuts & Bolts 6: *Kanonenwagen (SdKfz 251/9)* (Nuts & Bolts, 1997)

Web pages/Magazines
Hanomag SdKfz 251
http://www.geocities.com/MotorCity/Pit/3515/251/index.htm
Webmaster, Piet Van Hees
German SdKfz 251 Semi-track Part 1
http://www.kithobbyist.com/AFVInteriors/251/251a.html
German SdKfz 251 Semi-track Part 2
http://www.kithobbyist.com/AFVInteriors/251/251b.html
Webmaster, Mike Kendall
http://www.wwiitechpubs.info/garage/afv-deutschland/afv-de-ht-sdkfz-251/afv-de-ht-sdkfz-251-br.html
Webmaster, Justin Rigger

Museums and collections in possession of the SdKfz 251

Museum name	Production variant
Polish Army Museum, Warsaw, Poland	SdKfz 251/1 Ausf. D
Museum in Tomaszow Mazoweicki, Poland	SdKfz 251/1 Ausf. D
Victory Memorial Museum, Belgium	SdKfz 251/1 Ausf. D
Panzer Museum, Munster, Germany	SdKfz 251/1 Ausf. D
Patton Museum of Calvary and Armor, Fort Knox, U.S.A.	SdKfz 251/9 Ausf. D
The Tank Museum of the Royal Armoured Corps and the Royal Tank Regiment, Bovington, Great Britain	SdKfz 251/1 Ausf. C
Military Vehicle Technology Foundation, U.S.A.	SdKfz 251/1 Ausf. D
Musée des Blindés, Saumur-France	SdKfz 251/1 Ausf. D

Kits available

Considering the ubiquitous nature of the SdKfz 251, the representation of the vehicle in kit form has been very limited. Unfortunately for modelers, the quality of the majority of the kits listed below is considered poor by today's standards. On the other hand, at the time this book is being written, several companies such as Dragon and AFV Club have announced new versions of this venerable warhorse.

1:35 251/1/B/W Nitto. Circa 1965–70
1:35 251/1/C/R Tamiya
1:35 251/1/D/W Tamiya
1:35 251/9/D Tamiya
1:35 251/1C Dragon
1:35 251/1 D AFV Club
1:35 251/9 D AFV Club
1:48 251/1/B Bandai
1:76 251/1/B Matchbox
1:76 251/1/C Airfix.
1:76 251/1/C Fujimi
1:76 251/10/C Fujimi
1:72 251/1/D Hasegawa
1:72 251/22 D Hasegawa
1:72 251/16 Revell/Esci
1:72 251/10 Revell/Esci

Aftermarket and photo-etch sets

Eduard photo-etched set
35113 SdKfz 251/1 Ausf. D – general details
351 SdKfz 251/1 Ausf. C – general details

Part photo-etched sets
P35 035 SdKfz 251 Ausf. D – general details
P35 036 SdKfz 251 Ausf. D – floor, boxes and seats
P35 037 SdKfz 251 Ausf. D – stowage bins
P35 038 SdKfz 251 Ausf. D – fenders
P35 040 Mtl.SPW.SdKfz251/1 Ausf. D "Stuka Zu Fuss"
P35 041 SdKfz 251/3-IV Ausf. D "Rosi"
P35 042 SdKfz 251/1 Ausf. D (back doors)

Show Modeling
SM 010 SdKfz 251 D details

Moskit
3517 Muffler & exhaust

MR Models
3531 Floor/seat lockers
32120 Front wheels/spare wheels
SP-2 Torsion arms

Royal Model
051 SdKfz. 251 C photo-etch and resin details

086 SdKfz 251 D photo-etch and resin details
240 SdKfz 251 D (part 2) photo-etch and resin details
242 SdKfz 251 C/D interior plates
370 SdKfz 251 C Part 1
371 SdKfz 251 C Part 2
372 SdKfz 251 C Part 3
373 SdKfz 251 C/D wheels
376 SdKfz 251 C canvas cover no. 1
377 SdKfz 251 C canvas cover no. 2
383 SdKfz 251 C 251/7 Ausf. C no. 1

Tracks
Accurate Armor
SdKfz 251 track

Friulmodellismo
ATL-07 SdKfz 251 track
ATL-61 SdKfz 251 late track

Modelkasten
MK-19 SdKfz 251 track

AFV Club
AF 35043 SdKfz 251 Wheels and Tracks

Conversion sets
ADV/Azimut
35078 SdKfz 251 w/2cm flak (in actuality this is a SdKfz 11 with what resembles a 251 front end)
35120 SdKfz 251/2 8cm GRW34 mortar carrier conversion
35121 SdKfz 251/7 engineer vehicle conversion
35139 SdKfz 251/16 flamethrower conversion

R&J Enterprises
35251 Maybach HL42 TUKRM 6 cyl engine
35252 Front end with open engine compartment, one-piece or two-piece door options.
35252 Combination of 35251 & 35252 plus engine compartment details.
35008 Uhu IR searchlight conversion (SdKfz 251/20)

Chesapeake Model Design
CMD-31 SdKfz 251/22 conversion

Des Kits
35004 SdKfz 251/22 interior w/ Pak 40 gun conversion
35005 SdKfz 251/2 interior conversion
35049 SdKfz 251/17 2cm flak conversion

MR Models
42 SdKfz 251/7 engineer conversion
52 SdKfz 251/8 ambulance conversion
60 SdKfz 251/9 Kanonwagen Late conversion
62 SdKfz 251/22 Pak 40 conversion
3532 SdKfz 251/ final engine bonnet

Verlinden Productions
0739 SdKfz 251/7 assault bridge conversion
0564 SdKfz 251/9 Ausf. D update kit

Index

Figures in **bold** refer to illustrations

accessories and personal gear 5, 52, **52**, **60**
    painting **35**, **42**
AFV Club accessories 29, 47, **51**, 52, 53
airbrushes 8–10, **12**, **13**
ammunition belts **64**
ammunition cans **5**, 34, **37**
ammunition rounds and racks 34, **36**
    painting 34–35, **42**
Archer Fine Transfers 57, **57**, **68**
armor **33**, 34
assault bridges **53**
axels and wheel springs **51**, 52

bases 66, 67–73, **69**–**76**
Bego accessories 35, **43**, **76**
bolt detail **23**, 25
brackets 34, 53, **55**
buckets 35, **43**, **62**–**63**

carts, infantry **76**
compressors 10–11, **12**
cutting tools **9**

Decal Star accessories **39**
decals *see* markings and insignia
dioramas 66–77
doors
    crew **61**
    rear **28**, 53
Dragon accessories **64**, **76**
Dragon kits, modeling 21–31
drill bits **9**
drivers' and radio operators' compartments 6, 22, **24**, 34, **36**, 52–53
drives 21, **23**, 55

Eduard markings and insignia sets 35, **38**
engines and compartments 4, 5, 50–52, **51**, **60**, **61**
    painting **56**
fenders **23**, 25, **49**, 53, **55**
fighting compartments 5
    SdKfz 251/1 C 23, **24**
    SdKfz 251/1 D **20**
    SdKfz 251/7 52–53, **52**, **60**
    SdKfz 251/9 32–34
figures 66, **66**–**68**, **73**–**76**
glue **10**
grass, static 69–70, **69**–**71**
guns and gun mounts 32, **33**, **37**
    painting 35, **43**
    see also machine guns
handles 53, **60**
Hasawaga kits, modeling 16–20
hull floors **36**
hull interiors **22**, 23
hull roofs **56**
hulls, joining upper and lower 24, 26, 34, 35, **37**, 59

hulls, upper 22–23, **22**
Huntley, Ian 15

jerry cans and holders 29, **76**
Jordi Rubio width indicators **53**

Kubelwagens 67, **68**, 73, **73**–**76**

liquid cement **9**

machine guns **43**, 55, **64**
    painting 26, **29**
markings and insignia
    SdKfz 251/1 C **26**, **27**
    SdKfz 251/1 D **17**, 20
    SdKfz 251/7 **57**, **57**
    SdKfz 251/9 35, **38**, **39**
modeling tape **10**
Modelkasten tracks 32, 47, **53**, **57**
MR Models accessories 46–47, 49, **52**, 53, **55**
mufflers 34, **37**, **55**

New Connection accessories 34, **36**
Nicholls, Marcus 41, 42

painting and weathering
    chipping **41**, 42, 59
    color accuracy and scale effect 15
    detail washes 16–17, 26, **27**, 56
    finishing "style" 13–14
    "flattening" paint layers 17, **25**, **28**
    highlighting **25**, 26, **38**
    mud and grass effect 39–**41**, 40–42, **59**
    pastel weathering 20, **28**
    scribble method 56
    style of original 14
    techniques 12–13
    traditional washes 26
    weathering **27**
    weathering strategies 14
painting and weathering: specific features
    accessories 35, **42**
    ammunition rounds 34–35, **42**
    bases 67–70, **69**–**72**
    engines and compartments 56
    guns 35, **43**
    interiors **24**, 25–26, 34–35
    machine guns 26, **29**
    mufflers 34, **37**
    road wheels **17**, **59**
    tracks **58**, 59
painting and weathering: specific models
    SdKfz 251/1 C 25–29
    SdKfz 251/1 D 16–17
    SdKfz 251/7 55–59
    SdKfz 251/9 34–43
paints and pastels 11–13, **11**, **15**
pedals 34, **36**
photo-etch bending brakes **9**, 47, **48**
photo-etch detail assembly **46**, 47–48, **47**, **62**
photo-etch rolling tools **10**, **62**
planks, wooden 53, **53**, **55**

Plus Models accessories 35, **37**, **43**, **52**, **62**
punch and die sets **10**

R&J Enterprises engines **51**, 52
radios 35, **36**, **42**
road wheels 21, **23**, 53
    painting **17**, **59**
Royal Model accessories 29, **46**, 47, 53
running gear 34, 49

sanding tools **9**
SdKfz 251
    body types 4–5
    importance as modeling subject 7
    mission-oriented configurations 6
    origin and development 4
SdKfz 251/1 C
    description 5
    modeling 21–31
SdKfz 251/1 D
    description 5
    modeling 16–20
SdKfz 251/7
    description 6
    modeling 46–65
SdKfz 251/9
    description 6
    modeling 32–45
seats **22**, 34, **36**
soldering 48–49, **48**–**49**, **63**
springs
    door **61**
    wheel **51**, 52
storage racks and lockers 49–50, **50**, 53, **61**
suspension 21, **23**, 49

Tamiya kits
    251/1 D 32–65
    Kubelwagen **75**
tarps 35, **42**
tires, spare **29**
tracks 16, **19**, 21, 32, 47, **53**, **57**
    painting and weathering **58**, 59
trees **73**, **73**
Tristar figures 66, **76**

uniform patches **13**

Verlinden, François 66
Verlinden Productions accessories
    figures 66, **66**–**68**, **73**, **76**
    other accessories **51**, 52, **52**, **53**
vision ports and blocks **22**, 34, 52–53, **56**
visors 52–3, **53**, **56**

water, standing 71–73, **71**, **72**
water cans 35, **43**
water tanks **23**, **24**
weathering *see* painting and weathering
wheels *see* road wheels; springs, wheel
width indicators **53**
wire and wiring **11**, 25